The International Library of Philosophy

THE NATURE OF LIFE

Founded by C. K. Ogden

The International Library of Philosophy

PHILOSOPHY OF SCIENCE
In 6 Volumes

I	Scientific Thought	*Broad*
II	The Doctrine of Signatures	*Buchanan*
III	The Metaphysical Foundations of Modern Physical Science	*Burtt*
IV	The Nature of Life	*Rignano*
V	Scientific Method	*Ritchie*
VI	Biological Principles	*Woodger*

THE NATURE OF LIFE

EUGENIO RIGNANO

ROUTLEDGE
Taylor & Francis Group

First published in 1930 by
Kegan Paul, Trench, Trubner & Co Ltd

Reprinted in 2000, 2001 (twice) by
Routledge

2 Park Square, Milton Park, Abingdon, Oxfordshire, OX14 4RN
270 Madison Avenue, New York NY 10016

Routledge is an imprint of the Taylor & Francis Group

First issued in paperback 2010

British Library Cataloguing in Publication Data
A CIP catalogue record for this book
is available from the British Library

The Nature of Life
ISBN 13: 978-0-415-22568-7 (hbk)
ISBN 13: 978-0-415-61420-7 (pbk)
Philosophy of Science: 6 Volumes
ISBN 13: 978-0-415-22578-6
The International Library of Philosophy: 56 Volumes
ISBN 13: 978-0-415-21803-0

CONTENTS

PART I

Purposiveness of Vital Phenomena

PART II

An Intermediate Solution between Vitalism and Mechanism

INTRODUCTION

Before venturing to pronounce any judgment upon the vexed question as to the nature of vital phenomena, which has been for so long the subject of dispute between vitalists and mechanists (and of which the last enquiry undertaken by *Scientia* has been the most important echo), it is essential to bring into clear relief a fundamental characteristic common to all these phenomena and including as particular instances all those other properties which at various times have been put forward as characteristics of the vital phenomena themselves. This fundamental characteristic is the purposive or teleological aspect of the most typical vital processes.

It is, indeed, far from easy to say in what the purposiveness or even merely the purposive aspect of a phenomenon consists. To obtain a definition we have inevitably to take man, the measure of all things, as a starting point. Man explains the conscious and voluntary actions of his fellows by comparing them with his own; and since he knows by introspection that his own tend to satisfy certain appetites and desires, that is to realize certain ends, he attributes similar purposes to those of others; and, thanks to this hypothesis, he is able to explain their conduct in the way which is for him of the greatest pragmatic utility.

These same needs and purposes have driven man to invent and construct a tremendous variety of 'artefacts' (houses, manufactured articles of all kinds, together with the machines necessary for their production, etc.) which make their satisfaction and realization possible. Every 'artefact' or every machine in the widest sense of the word, as well as each of its parts, has, therefore, *an end*, *'it serves a purpose'*; and the purposive explanation of the machine, supplementing and completing the purely causal explanation, simply consists in discovering the purpose either of the machine as a whole, or of some one of its parts, or of its different ways of functioning.

Before proceeding further, however, we must note that the great majority of man's actions, instead of tending towards the immediate satisfaction of actual needs, rather anticipate and prepare for their future satisfaction, or even for the future satisfaction of those which, though not felt at the moment, will, according to our calculations, make themselves felt in time to come. These actions, therefore, cannot be explained from a purposive standpoint except in so far as the distant object which they envisage is in evidence. In the same way, and this is a particular instance of the purposive anticipation to which we have just referred, no machine serves the purpose for which it has been invented *until it is finished.* Before this point has been reached the purposive explanation of a machine or of any of the parts under construction consists in discovering the purpose which either *will serve* when the whole is completed and working.

The concept 'explanation,' in so far as it is concerned with human actions and creations, includes, therefore, in addition to the purely causal element, the purposive or teleological element, which completes it and is, indeed, the most important part of it ; in other words, it includes the discovery of the end or purpose which the action or object to be explained serves or will eventually serve.

This purposive explanation, far from being in opposition to the causal explanation, completes it, as I have said before ; and, moreover, *it is a complement which is absolutely necessary,* if, when brought into contact with human behaviour and human creations, our need for understanding is to be completely satisfied.

On the contrary, everyone knows that we do not experience the need for any such purposive explanation in the case of the phenomena of the *natural* inorganic world, in so far as they have not been artificially transformed by man. A purely causal explanation is completely sufficient for all phenomena of this kind.

Now we see that even people of the most rigorously positive scientific cast of mind, when confronted with the most typical and fundamental manifestations of life, are driven just as imperatively as in the case of human behaviour and human creations (which, after all, fall as such into the category of vital phenomena or their more or less direct or indirect results), to call to their aid *purposive* explanations, that is, to supplement their purely causal explanation with one analogous in certain respects to that which they apply to the behaviour and artificial products of man.

It cannot be that this need which we feel for a purposive explanation of vital phenomena springs from an insufficient knowledge of biology, because we feel no similar need in the case of inorganic phenomena which may for a time, on account of their great complexity, defy all our efforts towards a causal explanation. Nor can it be that it springs from an error or illusion of perspective caused by the standpoint from which we observe vital phenomena, for it still persists even when we try most vigorously to adopt the same attitude as that which we assume before phenomena of a non-living nature. But whatever the case, the fact remains that in so far as living beings are concerned we are led to adopt a special attitude which differs from that which we adopt where non-living entities are concerned.

This fact is sufficient to justify the statement that the substantial difference which exists between the attitudes a single mind adopts according as it is observing, though always from the same positive scientific standpoint, living entities or non-living entities, must correspond to some essential difference between the two great categories of phenomena into which man has divided everything in nature which is open to his perception.

In the first part of this work we propose, therefore, to find out whether this essential difference really exists and, if so, in what it consists. In the nine chapters of which it is composed we shall, as rapidly as possible, and in accordance with their particular purposive aspect, pass in review the most characteristic vital phenomena, beginning with the simplest and passing on to the more complex, in order to illustrate the nature of their common fundamental characteristic of purposiveness. At the same time we shall try to prove that our hypothesis as to the nature of life, which we have developed in our preceding works *On the Inheritance of Acquired Characters* and *Biological Memory*, is fully adequate to explain these purposive manifestations.

In the second part we shall try to indicate the exact position which our basic biological conception holds between the two opposing theories of vitalism and mechanism which, till now, have held the field alone ; and, in addition, to show that it is of a strictly positive scientific nature and explains the purposiveness of life on causal and deterministic lines.

In conclusion we shall outline the general considerations which our conception of life suggests as to the place of man in the Universe and the supreme aim of human existence.

The fact that this English edition follows so closely upon the Italian, French and German ones encourages us to hope that it will also arouse in the English-speaking public an interest in the renewed effort to lift a little the thick veil which shrouds the great mystery of life.

E.R.

Milan, January, 1930.

PART ONE

Purposiveness of Vital Phenomena

THE NATURE OF LIFE

CHAPTER I

PURPOSIVENESS OF THE MOST ELEMENTARY PHYSIOLOGICAL PHENOMENA

ASSIMILATION AND METABOLISM

Assimilation a selection. Property of self-reconstruction possessed by living matter. Synthesizing mechanism only the disintegrating process reversed. Metabolism as a process in a stationary state. Dissimilating phase as impetus to the subsequent assimilating phase. Tendency to self-conservation, a property peculiar to living matter. Necessity of recognizing in life a synthesizing process *sui generis*. Nature of this process according to the hypothesis here advanced.

EVEN the phenomena of such elementary and general vital processes as assimilation and metabolism possess well-defined and distinctive characteristics by which they are clearly differentiated from all the phenomena of the inorganic world.

Assimilation is a *selection*. The living substance, which is constantly being broken down during the phase of functional activity and constantly being built up again during the subsequent phase of reparative rest, *selects* from the extremely complex mixture of chemical substances dissolved in the nutritive liquid just those compounds or radicals or fragments of atomic groups which will reconstruct it in the same specificity as before. The *selection* is made by a substance which is different from these compounds, radicals or fragments of atomic groups, and which is certainly not inert even in its state of functional rest. It is reminiscent in certain respects of Maxwell's "sorting demon," which selects out of all the molecules of a gas only those which enable it to reverse the normal process of degradation of energy without violating the fundamental principles of thermodynamics. And as *selection* the assimilative process has a marked purposive aspect.[1]

Attempts have been made to compare the assimilative process and the growth of living substance which sometimes follows with the growth of crystals in a salt solution. The two processes are, however, substantially different. In crystal formation we see a reciprocal attraction of molecules, all inert and all equal to one another, which merely leave their places in the solution when it has reached a certain degree of saturation and pile themselves one upon another; in other words, we see a simple process of *deposition*. In assimilation, on the other hand, a substance which is not inert selects certain atomic groups or radicals out of the large number present in the nutritive liquid; they are different from the substance in question, the degree of dilution is not a determining factor, and they are just those ones which enable it to return to the state existing before the dissimilative phase. We are dealing, therefore, not with a merely passive process of deposition, but with *an active process of synthesis and reconstruction* by means of materials selected *ad hoc* and different from the substance which they are to build up again. As Arthur Thomson rightly says: "Organic growth takes place at the expense of materials different, we may even go so far as to say radically different, from those of which the organism is composed; it involves an

[1] *Cf.*, J. Johnstone, *The Philosophy of Biology*, 1904, p. 333.

active assimilation rather than a passive juxtaposition, as in the growth of crystals."[1]

The living substance, partially destroyed in the preceding dissimilative phase, thus appears in the assimilative phase to "aim at," "to have the purpose of" repairing itself and regaining its former condition. We see nothing similar to this in the inorganic world, where chemical substances, when disintegrated or transformed into other compounds, *do not show the slightest tendency to reconstruct themselves through an active process of their own.*

Loeb himself recognizes that "the essential difference between living and non-living matter consists in this: the living cell synthetizes its own complicated specific material from indifferent or non-specific simple compounds of the surrounding medium, while the crystal simply adds the molecules found in its supersaturated solution. This synthetic power of transforming small 'building stones' into the complicated compounds specific for each organism is the 'secret of life,' or rather one of the secrets of life."[2] He also recognizes that "the constant synthesis of specific material from simple compounds of a non-specific character is the chief feature by which living matter differs from non-living matter."[3] He is thus forced into admitting the existence of a "specific synthetic mechanism" which varies specifically for the various cells of different tissues.

But this "synthetic mechanism," whose function consists not simply in effecting the synthesis of complex compounds, but in building up again exactly those

[1] J. A. Thomson, "Vitalisme méthodologique," *Scientia*, January, 1923, pp. 26–27.

[2] Loeb, *The Organism as a Whole*, 1916, p. 23.

[3] Loeb, *ibid.*, p. 29.

compounds which were destroyed in the immediately preceding dissimilative phase, *must necessarily depend on the specificity and the modalities of this same disintegrating process.* The dissimilating process varies with specifically different external stimuli and consequently the quality or specificity of the part of the substance which has to be reconstituted varies accordingly. Corresponding changes must take place, therefore, in the synthesizing mechanism, *so that each time it reconstructs exactly and exclusively the self same substance which was destroyed during the particular process of dissimilation which it has to repair.*

The synthesizing mechanism, in other words, must bear the same relation to the disintegrating process as a medal to its die. When the die changes, the figure on the medal changes in exact correspondence; when the disintegrating process changes, the integrating mechanism must change in exact correspondence.

It is as if the disintegrating process itself were reversed, while remaining identical in every other way, and transformed into the corresponding synthesizing mechanism. Nothing comparable to this is seen in inorganic nature, where every process of composition or synthesis depends solely on the chemical properties of the materials in contact and on their physical environment, but *never on the character of the preceding process of disintegration which produced these materials.*

This is, then, an essential characteristic of assimilation, the fundamental process of life, and one which clearly differentiates the vital process from every purely physico-chemical process.

Passing from simple assimilation to that process of alternate assimilation and dissimilation which we call metabolism, we meet with another property of life which is never exhibited by any process in the *natural* inorganic world (that is, apart from man-made machines) : namely, the ability of this metabolic process to maintain itself in or settle back *into a state of ' stationary ' equilibrium* (to use Ostwald's terminology).

According to Hering, assimilation and dissimilation are the essence of life : " The essential characteristic which differentiates living matter, considered from a physiological point of view, from dead matter, is to be found in its organic exchanges ; that is, in the chemical processes taking place within the substance, which result, on the one hand, in the formation of materials which are excreted as foreign bodies—these materials being either accumulated in its immediate neighbourhood or sent out into the circulating fluids—and, on the other hand, in the absorption by the living substance, at the same time and by means of the same processes, of the nutritive materials which it assimilates and makes an integral part of itself."[1]

But most characteristic of the phenomenon is the fact that the " chemical processes which take place within the living substance " maintain themselves or tend to maintain themselves in a state of dynamic stationary equilibrium ; in other words, as Hering says, assimilation and dissimilation " balance each other, so that the living substance remains unchanged; both qualitatively and quantitatively."[2]

Piéron remarks upon the same fact, though with less precision than might be wished for : " Everything takes place in organic matter as though the preservation of an unstable chemical equilibrium were its inherent purpose."[3]

[1] E. Hering, *Zur Theorie der Vorgänge in der lebendigen Substanz.* Leipzig, 1921, pp. 53–54.
[2] Hering, *ibid.*, pp. 54–55.
[3] H. Piéron, " Du rôle et de la signification du conflit scientifique entre mécanisme et vitalisme," *Scientia*, February, 1922, p. 124.

When Thomson says that the primary fundamental characteristic *of the organism* is its "power of persisting in a specific complex metabolism," he is inverting the true order. This property belongs *not to the organism* but *to metabolism itself*, and it is one *not only of persisting in a specific complex metabolism*, but *of maintaining or striving to maintain this metabolism in a state of stationary dynamic equilibrium;* while it is from this inherent property of metabolism that the organism itself derives its most characteristic properties.[1]

It is not, then, the quality or kind of chemical transformations that take place in metabolism, nor the fact that they occur otherwise than in the laboratory (for example, at low rather than at high temperatures) which differentiates metabolism from every physico-chemical process in the inorganic world; but rather that it is essentially a *process in a dynamic stationary state*; that it does not take place except in a stationary state; and that if this state is interfered with by disturbing agents it successfully strives (as we shall see later when considering the phenomena of new adaptation) to eliminate or neutralize the disturbing cause, or else to involve it in its own process, and thus settle back into the former stationary equilibrium or into a new stationary equilibrium. Now in inorganic nature, until it is transformed by man's work, *we never meet any process with a similar tendency to maintain itself in or settle back into a state of dynamic stationary equilibrium, nor do we ever meet any process that is in such a state.* The engine which moves the machinery of a factory is, it is true, a mechanical system in a state of dynamic stationary equilibrium, that is, its movements, its dynamic modalities, do not vary with time; but the engineer knows what a number of inventions of valves to

[1] J. A. Thomson, *art. cit.*, p. 26.

regulate the pressure in the boiler, of centrifugal governors to control the distribution of steam, of heavy flywheels to equalize by their inertia the movement given to them by the piston, were required before the uniformity of movement necessary for the proper working of the machinery could be realized. In the same way the water which flows to a turbine is in a state of stationary hydrodynamic equilibrium, but only as a result of appropriate valves which keep the current constant. The constant pressure boiler is likewise in a state of thermodynamic stationary equilibrium, but only because of the safety-valves which preserve a steady constancy of pressure. So, too, the flame of a candle, to which many would liken the torch of life, is maintained in a state of chemico-dynamic stationary equilibrium, but only thanks to the wick, which man has constructed in such a way that there is a constant flow of melted wax through it.

There is then no need to cite any special vital process in order to show that the physico-chemist has so far failed to reproduce it in his laboratories or to explain it in purely physico-chemical terms; on the contrary, it is the most elementary and universal life process, viz., metabolism, which, by this essential property of being a process of a stationary nature, presents itself as fundamentally different from any process in the *natural* inorganic world.

It may perhaps be objected that organic growth, as such, seems to run counter to this property assumed for metabolism, since by virtue of it the living substance would neither increase nor diminish, but would always remain the same, both in quality and in quantity, during the constant alternations of assimilation and dissimilation. But organic growth, which is, after all, not a permanent but a transitory phase of life, is not due simply and solely to metabolism, but either to the presence of special internal energetic accumulations which at this point become active again, or to the stimulus provided by external energetic factors, especially thermic factors, which, by transformation of energy, give rise to new quantities of vital activity. What is important, and confirms the rule, is the fact that these new quantities of vital activity immediately exhibit a tendency to enter a stable metabolic state. Cellular division, for example, is one of the most typical manifestations of this tendency; for, as Spencer pointed out, it tends to restore the requisite proportions between the volume and the external surface of the cell, proportions which organic growth has disturbed and which are necessary for the maintenance of the normal metabolic state.

To sum up: the purposive aspect of metabolism is evident, first, *in its tendency to remedy immediately the organic destruction caused by dissimilation* by a corresponding assimilation which reproduces identically the part destroyed, both qualitatively and quantitatively; secondly, *in its tendency to remain in a stationary state*, different from any in the natural inorganic world, but similar to that of many machines or artificial processes to which man gives dynamic uniformity for his own ends; and thirdly, *in the tendency or property of self-conservation* with which it thus supplies every fragment of life and the organism as a whole.

All the tissues, for example, remain the same, although, as Johnstone says, " the material substances of which they are composed are in a state of continual flux."[1] And, in the same way, the organism as a whole, as Thomson points out, " preserves its integrity for days or months or years or centuries, always burning but never being wasted away."[2]

[1] Johnstone, *op. cit.*, p. 37.
[2] J. A. Thomson, *art. cit.*, p. 26.

"Living organisms," writes Loeb, "may be defined as *chemical machines* consisting chiefly of colloidal material and possessing the peculiarity of preserving and reproducing themselves."[1] But in the whole inorganic world, before man transforms it, there is nothing remotely resembling a chemical process tending to preserve itself, while the very use of the word 'machine' is a tacit recognition of the fact that we are dealing with properties which no physico-chemical process in the natural inorganic world exhibits.

The most casual observer must be struck by the difference between the living human organism, which remains almost unchanged during the thirty years of maturity between the ages of twenty-five and fifty-five, and the dead body which suffers such a complete change in less than twenty-four hours. "The dead body," writes Loeb, "if its temperature is not too low and if it contains enough water, undergoes rapid disintegration. It was natural to argue that life is that which resists this tendency to disintegration. . . . In the definition of Bichat, 'life is the sum total of forces which resist death.'"[2]

Why is there so substantial a difference in behaviour if only the same physico-chemical agents are active in the living organism as those which cause the rapid disintegration of the corpse?

And what explanation is there for the fact that the protein of the living cells of the walls of the stomach or the intestine is not digested by the digestive juices secreted by these organs in the same way as the protein ingested in the form of food? "The hydrochloric acid and the pepsin in the stomach and the trypsin in the intestine digest proteins taken in the form of food; why do they

[1] Loeb, *op. cit.*, p. 128.
[2] Loeb, *ibid.*, p. 349.

not digest the proteins of the cells of the stomach and the intestine? They will promptly digest the stomach as soon as the individual is dead, but not during life."[1]

Loeb none the less uncompromisingly rejects "*the monstrous assumption*" of a distinction between a "living" molecule of proteins and a "dead" molecule. According to him "the difference between life and death is not one between living and dead molecules, but more likely between the excess of synthetic over hydrolytic processes."[2]

But Loeb does not attempt to explain why, in the same molecule, there is an excess of synthetic over hydrolytic processes so long as it is living and the reverse as soon as it is dead. The explanation which immediately suggests itself is that life is *an energetic synthesizing process*, due, so to speak, to a swinging or whirling movement which carries the substance which we call living along with it, and which, when the process stops, leaves this same substance to the mercy of natural physico-chemical agents which are chiefly of a disintegrating character. In short, the process would be somewhat akin to those of water-spouts or whirlwinds which raise up drops of water or grains of sand with tremendous force and then abandon them to the action of gravity after their transformation into other energetic processes has caused a cessation of the whirling movement.

If this is so, the substance produced during the assimilative phase of metabolism is only the final chemical result of an energetic process *sui generis*, which is not merely of a chemical nature and which constitutes life.

"Even if we accept as incontestable truths," writes Gredilla the botanist, "the different hypotheses by means of which we seek to explain the successive formation, *in vivo*, of carbohydrates and other ternary products,

[1] Loeb, *ibid.*, p. 350. [2] *Id.*, *ibid.*, p. 354.

fats, and phosphorus molecules up to albuminoid substances, there still remains the great and difficult question: How does albuminoid matter acquire that instability which characterizes living matter?"[1]

Now in the light of the analogy we have suggested the instability which is typical of living matter arises out of this energetic process *sui generis,* which catches up and carries the albuminoid matter along with it during the metabolic phases both of dissimilation and assimilation.

We have thus considered, though only cursorily, the following facts:

1. The aspect or appearance of *selection* exhibited by assimilation.

2. The property of *spontaneous self-reconstruction* which, thanks to this assimilation, the living substance displays.

3. The synthesizing mechanism, *which seems to be simply the process of disintegration reversed* and which bears the same relation to it as a medal to its die: that is, it is as if the disintegrating mechanism itself changed direction, while remaining identical in every other way, and transformed itself into the corresponding synthesizing mechanism.

4. Metabolism, which appears as *a process in a state of dynamic stationary equilibrium.*

5. The dissimilative phase, representing in itself the impulse to the subsequent assimilative phase; this has already been noticed by Claude Bernard, and it is proved by the rapidity with which metabolism tends to repair the organic destruction which constitutes functional activity.

[1] A. F. Gredilla, "L'assimilation est-elle une fonction purement chimique?" *Scientia,* November, 1917, pp. 348–349.

6. The consequent tendency to self-conservation which is a characteristic peculiar to living matter.

7. The appearance of life *as an energetic synthesizing process sui generis* which reconstructs the living substance by the very fact of drawing it into its orbit.

In the presence of these facts, which so clearly differentiate the phenomenon of life, even in its most general and elementary manifestations, from all the phenomena of the natural inorganic world, we must ask whether it is in keeping with a true scientific spirit to persist in denying that we have here something substantially different from ordinary phenomena of a purely physico-chemical nature, and to continue to contend that physico-chemical laws alone afford a sufficient explanation. Is it not more in accord with sound positive method to enquire whether, by assuming a new form of energy—always obedient to the general laws of energetics, but endowed with well-defined elementary properties different from those of any form of energy in the inorganic world—it may not be possible to 'explain' these characteristic properties of life, which seem so many 'mysteries' simply because we obstinately try to explain them by physico-chemical laws which prove to be inadequate?

This is what we have tried to do, in our hypothesis concerning the nature of the vital process, expounded in the *Inheritance of Acquired Characters* and *Biological Memory*. We propose to give a rapid résumé of it here.

According to the theory of energetics, the various forms of energy are the products of two factors, of which one expresses a capacity and the other an intensity or potential. Thus the *intensity* factor in electrical energy is given by the potential or electro-motive force, and the *capacity* factor by the quantity of electricity. We assume to-day, as everyone knows, that this last factor is composed of elements termed 'electrons,' all of which are of the same capacity for all possible currents. In mechanical energy, on the other hand, we assume that the *capacity* factor, represented by mass, is made up of elements termed 'molecular masses,' which are specific for each substance—that is to say, they are of different capacity in different substances.

Now if we suppose that the *capacity* factor of this new form of energy which we need to postulate in order to explain vital phenomena (and which, for reasons to be stated later, we shall call nervous energy), is likewise sub-divided into elements which, on the analogy of the electrons of currents of electricity, we can call '*nervions*,' but which, differently from the electrons, are *specific* for each nervous discharge or current—that is to say of different elementary capacity in different currents—then we can assume a reciprocally univocal correspondence between specificity of discharge and specificity of substance.

It is easy to imagine that two molecules of different composition and structure could, by their sudden decomposition, give rise at the same time to different products which would represent the biochemical function properly so-called, and also to energetical nervous shocks of different capacities; and, on the other hand, it is conceivable that this process might be reversible: that is, if the same energetical nervous shock took place in the opposite direction it might succeed in rebuilding the structure which had been broken down. This is the more probable since, in the case we are considering, it would probably not be a case of complete destruction, which could only be repaired by a reconstruction from the foundations, but merely of the separation of a given lateral group of atoms previously united to the principal central mass by the interposition of the imine radicle NH, or the amine radicle NH_2, or the carboxyl radicle COOH, or some other similar radicle. Hence it would simply be a matter of re-attaching the lateral group to the central residue.

Thus Verworn, in his 'biogen' hypothesis, supposes that in dissimilation only the non-nitrogenous groups of atoms are detached, whilst the nitrogenous groups remain and constitute the central mass. This central mass later restores its original labile complexity at the expense of new non-nitrogenous atomic groups, similar to those cast off, which are contained in the surrounding nutritive liquid.[1]

If then all sorts of lateral groups of atoms are present, ready-made in the nutritive fluid, we may well assume that when an energetical shock of the same energetical 'capacity' as that previously produced by the detachment of a given lateral atomic group, of a well-defined quantity of mass, is again propagated through the nutritive fluids, it sets in vibration only this same atomic group, selecting it as it were from amongst all the others which are present with it in the liquid, and thus causing its re-union with the central mass.

The facts of fertilization and conjugation in general, moreover, seem reducible in essence to a simple 'coupling' or *placing face to face* of elementary quantities of living substance whose qualitative equivalence is proved by the fact that the male and female nuclei and the respective chromosomes taking part in the 'coupling' possess the same capacity for transmitting hereditary qualities. Similar proof is furnished by experiments on the substitution of the nucleus of a spermatozoon for the nucleus of an egg, and on the self-fertilization of the half nucleus of an egg by its other half, which had been expelled after the reducing division, and by many other experiments of the same kind. Similarly the strange nuclear processes

[1] *Cf.* M. Verworn, *Die Biogenhypothese*, Jena, 1923, pp. 35–46.

of synapsis, with which maturation begins both in the egg and in the spermatozoon, brings to view extremely delicate threads like necklaces arranged in pairs parallel to one another, so that each minute granule of chromatin of one filament is opposed to a similar granule of the other filament. And a similar coupling of granules of chromatin is seen in many other nuclear phases, both of germinal and somatic cells. These facts suggest that it is between opposed couples of qualitatively identical chromatin elements that the production of the vital phenomenon properly so-called takes place.

If we now assume that between each couple of these elements an intra-nuclear oscillating nervous discharge is produced, similar in certain respects to the oscillating electrical discharges of Hertz's resonators, we get a first rough idea of the inner nature of the vital process which we can submit to the test of facts.

In this connection the observation made by Engelmann will be remembered: namely, that the colours in the spectrum which are most readily absorbed by bacteria are those most favourable to their metabolism. *This metabolism would thus appear to be a phenomenon of a vibratory nature capable of undergoing resonance.* We may infer that just as the oscillating electrical discharges are kept from extinction, or rather reinforced by the synchronous Hertzian waves, so the oscillating intra-nuclear discharges of nervous energy may be kept from extinction or even revived by synchronous oscillations of light and heat, which are themselves Hertzian waves only of much shorter length.

We should thus obtain, without requiring to look for other causes, an explanation of the renewal and revival of life with the first coming of the warmth of spring.

But we wish above all to emphasize the fact that this

rough hypothesis concerning the inner nature of the vital process offers a ready explanation for all the characteristic properties of assimilation and metabolism which we discussed above.

The reciprocally univocal correspondence between specificity of living substance, which is alternately broken down and built up again, and specificity of nervous energy in its alternate phases of discharge and charge, which, as we have seen, arises from the strict correspondence between their specific energetic capacities, would thus be equivalent to a *selection* and would offer a very simple explanation of the *selective process* which takes place in the nutritive fluid and which many authors have regarded as the peculiar characteristic of assimilation.

At the same time we should find in the oscillating intranuclear discharge of nervous energy an explanation of the property of *spontaneous self-reconstruction* which we mentioned above. The specific nervous discharge produced by the living substance during its disintegration in the first half of the oscillating phase, would in the second half be transformed each time into a charging current capable of exactly reconstructing the same substance whose previous disintegration produced this discharging current. Hence the tendency towards *self-conservation* which is the exclusive property of living substance. Thus, as we saw that the facts themselves suggest, the synthesizing mechanism would be *merely the disintegrating process reversed.*

The dissimilative phase would thus be, as Claude Bernard held, the *impulse to the subsequent assimilative phase ;* and the tendency which metabolism shows to place itself and remain in a state of stationary equilibrium would result from the fact that the discharge of nervous energy caused by the disintegration of the living substance would immediately be converted into a charging current which would reproduce in identical quality and quantity the living substance destroyed in the immediately preceding phase. Further, the living substance would owe to this oscillating process, which carries it along in its own rhythm and rebuilds it at each new beat, its ability to persist and to resist the physico-chemical forces of dissolution—an ability which would be lost as soon as the oscillating process ceased.

We cannot stop here to show how this oscillating intranuclear metabolic process which constitutes, in its double phase of dissimilation and assimilation, the state of restorative functional rest, alternates with non-oscillating discharges of nervous energy coming from the nucleus itself, proceeding from the nucleus to the protoplasm and constituting disintegrating functional activity properly so-called; nor how this exclusively disintegrating functional activity goes on to exercise, as a result of the stronger impulse which it gives to the oscillating intranuclear process, a trophic action on the particular cells which are the seat of this functional activity. Nor can we stop to show how from this hypothesis an explanation of the rejuvenating influence exerted by fertilization can be deduced: a characteristic and exclusive phenomenon of life which has no analogy whatsoever in the inorganic world and which still remains, in spite of experiments on artificial fertilization which is merely stimulated parthenogenesis, a manifestation peculiar to life.

We must also leave unelaborated the explanation which this hypothesis affords of the reducing division which precedes fertilization and of the possibility, demonstrated by Maupas in his experiments on infusoria, of replacing conjugation in unicellular organisms by an exceptional functional activity maintained by the variability of the medium.

For all these questions we must refer the reader to the works mentioned above, and, in particular, to Chapter IV of *Biological Memory*, which is devoted to an exposition of the energetic properties of the nervous energy which we have assumed as the basis of life.

Our object in giving this rapid and necessarily incomplete account of the purposiveness of the most elementary physiological phenomena and of the explanation which our hypothesis offers is to show that to persist in denying that life has peculiar characteristics of its own and to maintain that physico-chemical laws afford a sufficient explanation of it, involves shutting one's eyes to irrefutable facts and adopting an anti-scientific attitude. A prejudiced refusal to embark on new paths is proof of intellectual sterility; on the other hand, to seek new hypotheses regarding the energetic properties of some other form of energy postulated at the basis of life and to submit these new hypotheses to the test of facts is—even though these first attempts may achieve but little success—to follow those methods which have always resulted in the greatest and most fruitful conquests of human knowledge.

We shall be the better convinced of this the further we advance, in the following chapters, in the examination of the other purposive aspects of life which are the more or less direct consequences and the more or less complex manifestations of those general and elementary purposive aspects which we have here considered.

CHAPTER II

Purposiveness of Generative and Regenerative Phenomena

Power of self-construction possessed by organisms. Harmony of development. Predetermined end towards which ontogenetic development tends. Development appears to be guided or directed by a kind of occult intelligence or entelechy. Faculty possessed by embryo of reacting to early disturbing influences in such a way as to arrive at its predetermined end. Purposive aspect not the least pronounced of all the phenomena of regeneration. Alleged morphogenetic action of physico-chemical factors merely a release of pre-existing morphogenetic potentialities. Possibility of a causal and deterministic explanation of the purposive aspects of generation and regeneration by means of the centro-epigenetic hypothesis of development. Nature of this hypothesis.

The faculty of reproduction or self-construction possessed by organisms has always been regarded, even in popular opinion, as a fundamental characteristic of life, to which no phenomenon in the natural inorganic world has the remotest resemblance. Even the machines which man makes for his own ends lack this power, for no machine has in itself the capacity of producing its own organs and its own elements. Such, too, has always been the impression of eminent biologists who have not been blinded by the preconceived idea that physico-chemical laws can by themselves provide an adequate explanation of life.

"The uniqueness of the living machine," writes Conklin, "is nowhere more evident than in its capacity for reproduction."[1]

"The biologist," writes Grassi, "must ask himself this question: is it possible to imagine a machine as complex

[1] E. G. Conklin, *Problems of Organic Adaptation*, The Rice Institute, Houston (Texas), 1921, pp. 307.

as the living being which forms, repairs, recompletes and recomposes itself automatically, even when the order of its different parts is disturbed ?"[1]

"If we are going to compare the organism to a machine," says Stefani, "it must be to a machine which can make, preserve and reproduce itself, and which can, in addition, repair any breakages which may occur."[2]

Even if it were possible to explain the organism as being a physico-chemical machine, the most fundamental question would still remain unanswered—viz., how did this machine make itself ? As Kant says : "the machine has merely moving power ; but the organism has, in addition, formative power."[3]

The purposiveness of ontogenetic development is too evident to be denied seriously.

It results from the convergence of manifold morphogenetic activities towards one sole end, that is, towards the formation of a marvellous functional unity, every part of which serves to maintain the life and guarantee the well-being of the whole. We repeat that nothing even remotely analogous is to be seen in any physico-chemical process of the inorganic world. To quote Grassi again : "When a person considers the way in which a chicken springs from an egg he experiences a deep feeling of wonder ; the biologist is acquainted with the smallest details of this process of development, but this feeling is in no way diminished, for he sees a miracle take place before his eyes ; it is as if he should see a palace growing of its

[1] G. B. Grassi, *La vita. Ciò che sembra a un biologo.* Lecture given before the R. Accademia dei Lincei, June 3rd, 1906. Rome, 1906, p. 5.

[2] A. Stefani, *Sul concetto della vita.* Inaugural lecture given at the University of Padua, November 5th, 1906. Venice, 1908, p. 4.

[3] H. Driesch, *Der Vitalismus als Geschichte und als Lehre,* Leipzig, 1905, p. 69. (English Edition, *The History and Theory of Vitalism,* London, 1914, p. 74. Part II, to which reference is made *infra,* was rewritten by the author for this English edition ; subsequent references are therefore to the original.—Ed.)

own accord, without the help of workmen, out of a heap of bricks and plaster, with its windows and doors gradually opening, and its panes and shutters, its balconies and furniture taking shape."[1]

This "harmony of composition," as Driesch calls it, has indeed a touch of the marvellous. "Experiment shows," he writes, "that the organs of the adult which, in spite of their complexity, form an indivisible whole, arise from the re-union of several elements each of which develops independently. There must, therefore, be some harmony presiding over the relative position of the parts of the embryo, as a result of which the independent elements unite to form a whole. We call this harmony 'harmony of composition.' An excellent example of it is to be seen in the development of sea urchins, in which the mouth and the intestine develop independently, but when the development is completed the two organs fit together perfectly."[2]

Ontogenetic development evidently aims at a predetermined end: "In all of its general features," writes Conklin, "development is teleological. The end is apparently in view from the beginning. Consider, for example, the development of the eye: the retina with its sensory rods and cones, the lens with the ciliary processes and muscles for focussing, the transparent cornea and humor—each and every portion of the organ develops toward the end, or shall we say 'for the purpose' of vision, and yet there is no vision until after all these parts are formed and connections have been made with the central nervous system, which does not occur until late in development, sometimes, as in the case of the rat, some time after birth."[3]

[1] Grassi, *art. cit.*, p. 4.
[2] Driesch, *op. cit.*, p. 176.
[3] Conklin, *op. cit.*, pp. 315, 316, 329.

The most striking display of purposiveness in ontogenetic development lies in the fact that the organism forms in itself organs which will *later* serve for its conservation. The formative activity seems, in other words, to be endowed with a 'prevision' of future needs, as when, for example, to quote Conklin's illustration, it creates the organ of sight in the embryo while still in the maternal womb, although it will be of use to the individual only in its extra-uterine life; it creates the stomach and all the other parts of the digestive apparatus, useless to the embryo which is still nourished by the maternal blood, and the lungs, though no respiratory function is yet necessary.

Thus ontogenesis seems 'marshalled' by some occult intelligence or 'entelechy' in the same way as the construction of a machine and the arrangement of its different parts is presided over by the mind of the engineer. "If we watch the development of the egg of an animal into a larval form," Johnstone writes, "and continue to trace the metamorphosis of the larva into the perfect animal, we cannot fail to conclude that, beside the individual physico-chemical reactions which proceed, there is also *organization*. The elementary processes must be *integrated*. There must be a due order and succession in them. In studying developmental processes, in considering the developing organism as a whole, we are impressed above all else with the notion that not only do physico-chemical reactions occur, but that these are *marshalled* into place, so to speak."[1]

The purposiveness of ontogenetic development, that is the existence of this predetermined end, is also evidenced by the fact that the embryo overcomes early disturbances which might deflect it from its course. Such disturbances would certainly deflect a purely physico-chemical process;

[1] Johnstone, *op. cit.*, pp. 127–128.

but the embryo reacts against them so as to get back into its usual path and resume its normal development. "A change in the conditions under which the inorganic system differentiates," writes Johnstone once more, "leads of necessity to a different final phase, but a change in the conditions under which the embryo develops need have no such effect. If some unforeseen occurrence takes place—some artificial interference with the process of segmentation, which could never have been experienced in the racial history of the organism—a regulation by the parts of the embryo occurs, and the final phase of development may be the same as if no interference had been experienced."[1]

Monstrosities themselves are evidence of the tendency which the developing embryo shows to arrive at all costs at its predetermined end; for even the most serious accidents which can happen during the first ontogenetic phase at the most only *arrest* the development of a particular part at a particular point, whilst the remaining parts of the embryo adapt themselves as best they can and with as little modification as possible to these partial local arrests, and thus, in spite of everything, preserve their specific form with the least possible change.

In the phenomena of regeneration we see again this tendency of the organism as a whole, as well as of its parts, to arrive at any cost at its predetermined end or to return to the path leading to it when it has been diverted by some violent interference.

The teleological aspect of these phenomena appears still more pronounced when we consider the ways which the regenerative process follows—ways often different from ontogenetic paths, and chosen whatever they may be if only they will lead to the regeneration of the part

[1] Johnstone, *ibid.*, pp. 321–322.

amputated or the reshaping of the remaining part, so as to make a new complete organism. This is particularly noticeable in the phenomena of *Rückbildung* (regressions) or *Umdifferenzierung* (renewals of differentiation), in which the remaining part loses its differentiation, becomes a heap of undifferentiated cells, and begins anew the process of forming the new organism. Nothing similar happens in the reintegration of crystals, which ultra-mechanists, blinded by prejudice, have compared with the process of regeneration.

We will not speak here of the experiments made on the isolation or displacement of first blastomeres, in which exactly the same organism is produced as in the normal course of development, nor of the fusion of blastulæ and gastrulæ of the same kind with its production of a single organism. But these two series of experiments equally indicate the existence of a predetermined end towards which development moves, and from them Driesch has quite justifiably drawn his strongest arguments in support of his anti-mechanistic conception of life.

What do the ultra-mechanists do when confronted by these distinctly teleological manifestations of the generative and regenerative processes, which are clearly patent to the eyes of all who are not deliberately blinded by prejudice? They direct all their efforts to an attempt to prove that given chemical substances exercise a morphogenetic action on particular developments, hoping to conclude triumphantly that the entire series of morphogenetic phenomena constituting ontogenetic development may be explained completely and exclusively by physico-chemical action.

But in this attempt they have mistaken an action which merely *releases* morphogenetic activities for a genuine morphogenetic action. And besides, even if they succeeded in their task, they would still have to explain how this series of simple physico-chemical actions and reactions has assumed the purposive aspect which is so characteristic of developmental phenomena and which, on the other hand, is never seen in the physico-chemical processes of the natural inorganic world.

An examination of the facts which they bring forward in support of their theory will suffice to show that the ultra-mechanists have not succeeded in their attempt, and that they have confused actions of simple release with genuine morphogenetic actions.

Bohn, when trying to prove that only physico-chemical phenomena are involved in ontogenetic development, writes: "When fragments of the testis of newly caught frogs are grafted under the skin of castrated frogs, which have not, therefore, the sexual character of the swelling of the strongly pigmented thumb with its numerous glands, the glands of the thumb begin to develop again, and in less than two months they take on a normal appearance."[1]

"In his latest work," continues Bohn, "Steinach has experimented with cross graftings: he grafted ovaries on to castrated males and testes on to castrated females. Rats and guinea pigs were taken very young and the operation performed on all the animals of a litter. The first result to emerge was that the influence of the puberty gland, whether male or female, is specific. Not only does it not encourage the development of the organs of the opposite sex, *but it inhibits them*, and causes them to regress. On the other hand it causes the homologous sexual organs *to come out of their undeveloped and latent state*: in the castrated males which had ovaries grafted on, the mamillæ and milk glands grow until they are the same in form and size as in normal females of the same

[1] G. Bohn and A. Drzwina, *La chimie et la vie*, Paris, 1920, p. 126.

age. And in all other particulars they acquire female characteristics, both morphologically and psychologically."[1]

From these observations, especially from those which we have italicized, it is clear that the chemical substances or hormones secreted by the grafted ovaries and testes effect a simple release, a freeing of morphogenetic activities which had until then been suspended or inhibited. The phenomena of hermaphroditism offer abundant illustration of this, as Bohn himself seems to admit. "It is in insects," he writes, "that the most frequent examples of hermaphroditism are found, that is of individuals which are half male and half female, or a varied mosaic of the characteristics of the two sexes. *It is not easy at first glance to reconcile hermaphroditism with the hormone theory.* It is evident that the secretions both of the testis and the ovary, which are carried by the blood, do not remain in one half of the body: they circulate throughout the organism and influence the cells both of the male and the female half. Many biologists, therefore, are opposed to the theory of sexual hormones, as far, that is, as insects are concerned, for in vertebrates their action is too obvious to be denied. But examples of hermaphroditism are found in vertebrates also. There are, for instance, bullfinches which have on one side of the body the testis and male plumage and on the other an ovary and female plumage, the feathers meeting on the midline of the body without any transition. Is the hormone theory, therefore, inapplicable in this case? Not necessarily. A chemical substance which is dissolved in the blood, although it is in contact with all the tissues, acts only on those of correlative chemical structure, and not on the others."[2]

[1] G. Bohn and A. Drzwina, *ibid.*, pp. 130–131.
[2] *Id.*, *ibid.*, pp. 136–137.

That the chemical action of the hormones is nothing more than a releasing of morphogenetic activities and not itself morphogenetic, is equally shown in the opposite cases when the action of the hormones *inhibits* morphogenetic activities: "Péchard has demonstrated that the removal of the ovary leads to the appearance of spurs and male plumage in hens. The ovary would thus normally exercise an *inhibiting influence* on the development of these characters."[1]

An analogous action of simple release, which no one would take for a morphogenetic action, is exercised, for example, by water on the larvæ of the newt triton: "After having removed the larvæ of the newt triton at an early stage from the mother's body, Kammerer placed them in water, where they completed their development. In this case, they developed gills and other structures like those of their fishy ancestors."[2]

Many of these examples of simple release of morphogenetic and instinctive activities, effected by the external medium as a whole or by certain particular physicochemical stimuli, are described by Semon in his work *The Mneme*,[3] which deals with the analogy which exists between this release of morphogenetic or instinctive activities and the mnemonic recollection of certain mental activities. In a suggestive fashion Semon shows how many of the morphological passages from one stage of development to another are evoked by stimuli which, on account of their nature, or of the circumstances in which they work, could certainly never exert a true 'formative' action, but which may be easily

[1] G. Bohn and A. Drzwina, *ibid.*, pp. 137–138.

[2] C. J. Tomkeieff, "The Mnemic Theories of Evolution," *Scientia*, September, 1923, p. 166.

[3] R. Semon, *The Mneme*, London, 1921.

interpreted as effecting a release of certain morphogenetic tendencies.

In short, just as the smell of seaweed may evoke the image of the gulf of Naples in the mind of someone who has seen it before (and surely no one would maintain that the chemical action which is then exerted on the nasal mucus exerts at the same time a formative and directly determining influence in relation to the excitations of the cerebral cortex which constitute this image), so certain hormones are responsible for the release of a given morphogenetic activity. Thus, this type of argument which the mechanists resort to in no way proves that ontogenetic development can be produced by the simple play of physico-chemical factors.

But, we repeat, even if they succeeded in this attempt they would still have to explain the purposive aspect of this ontogenetic process, which is so complex and which has no analogy in any physico-chemical process of the inorganic world. Even if we admit, says Conklin, " that a given hormone is the chemical stimulus which leads to the adaptive reactions in each of these cases, it does not in the least explain the fact that these reactions are adaptive. Why should the reactions of so many different parts of the body to andrase (formed by the interstitial cells of the testes) or gynase (formed by the ovary) be of such a character that they lead to the development of all the complicated organization of the male or female, and why should the organization of the two sexes be so adapted to each other? It is evident that the stimulus which starts these adaptive reactions does not explain the fact that they are adaptive."[1]

[1] Conklin, *op. cit.*, p. 326.

In this section on generative and regenerative phenomena we have thus rapidly considered the following facts:

1. The developing organism is capable of self-construction.

2. All the morphogenetic activities converge towards a single end (' harmony of composition,' to use Driesch's phrase).

3. Ontogenetic development tends towards a predetermined terminal state.

4. The development of the embryo seems as if it were marshalled by a sort of occult intelligence or ' entelechy,' in the same way as the construction of a machine is directed by the mind of the engineer.

5. The embryo reacts against early disturbing influences in such a way as to resume at all costs its normal development and to arrive at its predetermined end.

6. The phenomena of regeneration have likewise a purposive aspect, indicating that the organism strives to return to its normal state.

7. Finally, the alleged morphogenetic action of physico-chemical factors is revealed as being a mere release of morphogenetic activities already potentially existing.

In view of all these facts by which the highly purposive generative and regenerative phenomena are so clearly distinguished from all the non-purposive phenomena of the natural inorganic world, it seems justifiable to ask once more whether it is in keeping with a true scientific spirit to close one's eyes to this purposiveness and to persist in affirming that there is nothing in these phenomena which cannot be explained by the common physico-chemical factors active in the inorganic world. On the contrary is it not in accordance with sound positive methods of research and investigation to consider the possibility of accounting for these phenomena of generation and regeneration by postulating a new form of energy—always obedient to the general laws of energetics, but endowed with well-defined elementary properties of its own —just as we have done in the case of the most elementary and general phenomena of life, viz., assimilation and metabolism? By renouncing a purely mechanistic or physico-chemical explanation, which has shown itself incapable of accounting for the most characteristic peculiarities of these phenomena in their purposive aspect, may it not be possible to explain them from the *energetic* and *deterministic* point of view, by the assumption of a new form of energy peculiar to the realm of life?

Now the property which we have assumed above for the new form of energy which we have postulated at the basis of life (which we shall continue to call 'nervous' energy for reasons to be given soon) furnishes us with the desired explanation. According to our hypothesis the specificity of nervous current or discharge which is produced by the sudden decomposition of a given nuclear substance is the same which, functioning later as current of charge, reproduces the same specific substance; or, inversely, every new nervous current, arising as a result of a transformation of energy from a given external stimulus or a new internal adaptation, deposits in the particular nucleus in which it ends the same specific nuclear substance which, in its subsequent decomposition, reproduces the same specific nervous current which deposited it, thus constituting a specific accumulation of this nervous current. This property of specific accumulation (or, in other words, of reciprocally univocal correspondence between the specificity of the nervous current, whether of charge or discharge, and the specificity of the substance accumulating this nervous energy) is more than sufficient, with the support of the centro-epigenetic hypothesis of development, to give us an energetic and deterministic explanation of all these generative and regenerative phenomena, even in their most clearly teleological aspects, without having need to resort to 'entelechies' or other metaphysical or mystical entities of the same kind, which, by their very nature, imply the abandonment of any scientific causal explanation.

In our work on *The Inheritance of Acquired Characters: Hypothesis of a Centro-epigenesis*, we attempted to develop this theory. According to our hypothesis, during the whole course of ontogenesis a 'plasmatic' activity radiates out from a central zone of development, constituted by the germ-plasm and situated in a given part of the organism. This part is always the same, both when the organism in its phylogenetic changes successively deposits specific accumulations in the germinal substance, each one corresponding to the physiological activity which at this point is the resultant and representative of the general physiological state of the particular phylogenetic stage, and when, on the other hand, the germinal substance itself successively activates the accumulations which have been deposited in it during phylogenetic evolution and thus produces the unbroken passage of the embryo from one ontogenetic stage to the next.

We have thus succeeded in solving the three fundamental dilemmas of ontogenetic development which have so long defied biologists, viz., preformation or epigenesis, preformistic germs or non-representative germinal substances, and nuclear somatization or qualitatively equal nuclear division. We have also been able to account for the fundamental biogenetic law of the ontogenetic recapitulation of phylogenesis; for the mechanism of the transmission of acquired characters; and for the notable analogy, brought out in particular by Hering and Semon, between the phenomena of development and mnemonic phenomena in the narrow sense.

As far as this last analogy is concerned, we have even been able to show that it resolves itself into a case of substantial identity, in that the phenomena of these two categories seem to be particular manifestations of the elementary and general fact of specific accumulation which is found at the basis of both.

A comparison of the phenomena of transmission of nuclear excitations along the intercellular protoplasmic filaments of tissues in both animals and vegetables and the phenomena of sensitiveness and movement which are seen in some degree in all plants, with the phenomena of movement and behaviour in general of unicellular organisms and of the embryonic forms of marine species which, even in the most primitive stages, are capable of an independent life and of movement, seems, on close examination, to justify the conclusion that the form of energy, which we have hypothetically assumed as the basis of life (and which we have provisionally called 'nervous' energy), is nothing other than nervous energy properly so-called, even when it is produced by the nuclear discharges of non-nervous cells, whether animal or vegetable, somatic or germinal. This conclusion, moreover, is in agreement with the well-known view held by Claude Bernard as to the essentially similar nature of all forms of irritability or organic excitation. And we can conclude unreservedly from this that all the specific accumulations deposited both in the nuclei of the very varied somatic cells of the different tissues and in the germinal cells are really authentic mnemonic accumulations, fundamentally of the same nature as those deposited in the centres of the grey matter of the brain, which produce by their discharge the psychic phenomena of mnemonic recollection in the strict sense.

For a detailed exposition of this hypothesis we must refer the reader to the work already quoted on *The Inheritance of Acquired Characters* (or to Chapters II and III of *Biological Memory*, where a résumé of it is given), for much ground still remains to be covered before we complete this examination, rapid though it is, of the principal purposive manifestations of life.

CHAPTER III

Purposiveness of the Phenomena of Pre-established Adaptation

Nothing similar to the phenomena of pre-adaptation to environment is ever seen in the natural inorganic world. Previous co-ordination existing between different parts of the organism has also a definitely purposive aspect. For every element or organ or function of the organism there is always an adequate reply to the question: of what use is (or has been in the past) this element or this organ or this function? Complete inadequacy of the Darwinian explanation of natural selection to account for this pre-established adaptation of organisms. Similarly, complete inadequacy of physico-chemical explanations which have recourse to enzymes, zymogens, zymoids, kinases, etc., to account for functional co-ordination. Passage by insensible degrees from purely physiological functionings to instincts in the proper sense. The only possible explanation of these phenomena of pre-established adaptation is to consider them as due to the hereditary transmission of new adaptations which have been produced in the past in the course of phylogenetic evolution. Necessity, therefore, of admitting the transmissibility of acquired characters, which the centro-epigenetic hypothesis accounts for, and of inquiring into the nature of the capacity for new adaptation which the living substance possesses.

As soon as the organism enters into contact with the external world, and before the latter has had any opportunity for exerting any formative influence upon it, it is already equipped with all the organs and already capable of all the functions and activities which fit it to its environment. In this pre-established adaptation we again find evidence of purposiveness—a purposiveness which is in strict dependence on the ontogenetic purposiveness already examined.

The pre-adaptation of every living species to its conditions of life is, as we know, one of the facts which has aroused the profoundest measure of wonder in naturalists, while it has inspired the religious to burst forth into rapturous hymns of praise to the author of a work revealing so remarkable a degree of foresight.

The whole complex organism, with its infinite variety of organs whose functions are all so remarkably connected and co-ordinated, and all of which 'serve' for the conservation and the well-being of the individual, is ready for embarking upon its independent life before this latter begins and, consequently, before the majority of its functions have had any opportunity of being exercised or of being useful to it. "Inherited adaptations," says Conklin, "are those which appear in the development of individuals as if in anticipation of future needs and not as a result of present ones."[1]

It is apparent even to the non-specialist that the stomach and the lungs of the mammal are already formed when it leaves the maternal womb, although only at this point do they begin to be of use for the maintenance of life; that similarly the chicken comes from the egg with its wings already formed, the new-born child has eyes ready to receive the first rays of the sun, its legs are furnished with ligaments and tendons and muscles as if they were there to invite the baby to make use of them, and teeth begin to appear while the child is still suckling.

But the biologist finds still deeper and more valid matter for astonishment in the further mechanism of complex co-ordinations whose network extends over the different parts of the organism and over the different parts of an organ, even before it has begun to function. "The whole sensorial and motor apparatus of the visual organ, in the widest sense of the word, is a living tool," Hering writes, "of which the new-born child has not laboriously to learn the use, but which rather procures for him, without effort or trouble, his first optical knowledge of the external world." "The sensory and motor processes of the eye come into action simultaneously after

[1] Conklin, *op. cit.*, p. 303.

birth, guiding each other in their functioning and assisting each other in the subsequent development of their innate power."[1]

Not less pronounced is the purposive aspect of all the varied means of attack and defence which organisms possess, means which are so perfectly suited to the purposes of the latter and which likewise develop before the animal has need of them and are thus ready for the first occasion which may arise. "When one considers," writes Conklin, "all these striking contrivances for defence and offence, together with the appropriate behaviour by which they are accompanied, such as the well-known habits of the rattlesnake, the porcupine, the opossum and the skunk, the question inevitably arises whether lower organisms have not discovered these means of protection in a manner comparable to the way in which man has discovered methods of defence and offence."[2]

In short, with regard to every organ, every function, every vital activity with which the organism is endowed from the very moment it begins life outside the egg or the maternal womb, the question: "Of what use is this organ, this function or this activity?"—a question which is in every way analogous to that which the engineer would ask about every element or part or function of a machine constructed by man for definite purposes—always receives an adequate reply. "With regard to every vital process," to quote Conklin once more, "we may properly ask the question, *cui bono*, being confident that it is or has been useful."[3]

How have mechanists attempted to explain this morphological adaptation, the purposiveness of which was too evident even for them to deny? They have

[1] Hering, *op. cit.*, pp. 133, 138.
[2] Conklin, *op. cit.*, p. 313.
[3] *Id.*, *ibid.*, p. 301.

thought to find the explanatory key in Darwin's natural selection, which owes its great and, originally, almost undisputed success to its attempt *to explain in a non-purposive way these purposive morphological phenomena of life.*

"It is no exaggeration to state," wrote the most extreme of modern mechanists, J. Loeb, who has recently died, "that the number of species existing to-day is only an extremely small fraction of those which can and possibly do originate, but which escape our notice and disappear because they cannot live or reproduce. If we consider these facts we realize that *the mere laws of chance* are adequate to account for the fact of the apparently purposeful adaptations."[1]

But the Darwinian explanation, which considered only the morphological aspect of pre-adaptation—the only one which at first attracted the attention of naturalists, since in it alone could they find scope for their classificatory work and since it was the one which Darwin had chiefly in mind when he worked out his theory—has been shown by a more careful examination, after the first wave of enthusiasm had evaporated, to be completely fallacious.

We cannot examine here all the evidence and all the arguments against the Darwinian explanation, which Weismann carried to its logical extreme. Without denying that the struggle for life and the natural selection which follows from it are responsible for elimination to a certain extent, these proofs and these arguments reveal the complete insufficiency of Darwinism as a single explanation of phylogenetic evolution. And when, above all, we are dealing with complex and co-ordinated morphological adaptations—by far the most numerous—which would involve the hazardous encounter of a large number of

[1] Loeb, *op. cit.*, p. 345.

particular modifications all fortuitously assembled at a given moment, in one and the same individual, the absolute insufficiency of this explanation becomes glaringly appaent.

These arguments and many others show, in short, the inconsistency of the Darwinian theory with regard to the chief factor of evolution which it assumes; in other words they show the complete impossibility of imputing the construction of an organism *to the mere play of fortuitous variations*, in which the organism would take no active part and which would come to it as a gift from heaven—even though this chance play were aided by an incessant, rigorous and most active selection—of entrusting it, as it were, to the casting of a dice, in which the organism would play only a passive part. For the construction of an organism means the construction of a mechanism more complicated and more perfect than our machines, one which answers a well-defined purpose, and is the complete antithesis of processes left to pure chance. But again on this point we must refer the reader to our work on *The Inheritance of Acquired Characters*, where all these proofs and arguments are put forward and discussed.

We must be content to consider here only a few points.

In the first place, only where life is involved has it been found necessary to resort to natural selection, that is to a process totally unknown in the inorganic world, where no phenomenon calls for any such explanation. This fact alone, therefore, leads us to place life in a category apart, distinct from that in which purely physico-chemical phenomena are ranged.

In the second place, to postulate natural selection is to recognize implicitly the tendency of life to endure and to propagate itself, to persist at all costs in its own energetic form, in such a way as to supply the sifting process of natural selection with abundant and constantly renewed materials. This, once more, is a characteristic typical of no form of inorganic energetic activity, each of which is produced by other forms of energy and transforms itself again into them, and none of which shows any tendency towards self-propagation or self-preservation, even when it is placed in an unfavourable environment.

The Darwinian theory, therefore, which mechanists have exalted on the score that it alone is capable of accounting for the morphological manifestations of the purposiveness of life without going outside the sphere of physico-chemical factors, on a closer examination and quite apart from its absolute insufficiency, is shown, on the contrary, to be wholly based on the presupposition of certain properties in life which, in themselves, differentiate it completely from all the phenomena of the inorganic world.

But, as we said above, the Darwinian theory primarily attempted to explain those phenomena of pre-established adaptation which are principally *morphological*. At that time there was little or no knowledge of *physiological* adaptation, which is even less explicable by the accumulation of purely fortuitous variations, even though sifted by natural selection. And in these physiological phenomena of pre-established adaptation the purposive aspect is, if possible, even more prominent. "The organism," writes Stefani, "defies explanation unless we recognize that it has the power to act and react in accordance with a purpose, a power which mechanism does not explain."[1]

Thus, for example, if the quantity of carbonic acid in the blood passes a certain limit, so that it would be harmful to the organism, special nervous cells situated in the respiratory centre react and excite the corresponding

[1] A. Stefani, *Sul concetto della vita*, p. 13.

respiratory muscles, the rhythm of respiration is accelerated, the respirations become deeper, and the excess of carbonic acid is expelled, so as to re-establish equilibrium in the blood. If, as a result of hæmorrhage, disease or any other cause, the red or white corpuscles disappear or are destroyed in great number, the organism provides for their replacement by producing a greater number than usual. No less perfect is the mechanism responsible for regulating the internal temperature of warm-blooded animals ; whatever the variations of the external temperature, the organism succeeds in keeping the fluctuations in its own temperature within the limits of some tenths of a degree.

" In these cases," continues Stefani, " what passes our understanding is not the simple anatomical or physiological facts considered in isolation—the excitation of a particular nerve, the contraction of a particular muscle, the secretion of a particular gland—but their co-ordination in the interests of the organism."[1]

Note, in addition, that no combination of physico-chemical processes in the *natural* inorganic world presents anything in the least like these physiological reactions, which enable the organism *to maintain itself in its stationary physiological state* in spite of variations which may take place in its environment ; on the contrary, every physico-chemical process varies always in perfect correspondence with its environment.

Nor is it only the higher organisms which possess this power of self-regulation ; it is seen in the lower organisms to no less a degree as, for example, in relation to osmotic pressure : " When the osmotic pressure of the environment changes," Driesch writes, " many aquatic organisms, algæ and fungi, for example, are capable of modifying

[1] Stefani, *ibid.*, p. 4.

by self-regulation their own internal pressure, either by changing the permeability of their skins to salts, or by forming within their cells substances which give rise to an increase or a decrease of their own pressure." "This power possessed by organisms of maintaining their normal configuration and functioning in spite of the influence of abnormal circumstances is always present to an astonishing degree."[1]

Here we must point out that, in the first place, our observations in the first chapter on the tendency of the elementary metabolic processes to maintain themselves in a stationary condition explain *ipso facto* the same tendency in the organism as a whole, since its entire activity results from the totality of the single metabolic cellular processes; and, in the second place, that this in no way diminishes the ulterior purposive aspect—in addition to that of the tendency in itself—of the various physiological procedures by which the organism succeeds in satisfying this tendency, inasmuch as the organism finds these procedures co-ordinated and ready to function as soon as occasion demands.

Still more characteristic are the other phenomena of pre-established physiological adaptation which appear in the form of *functional anticipation*, such as, for example, the swelling of the mammary glands in female mammals before parturition, so that they are ready to secrete milk as soon as the child is born. Even if we assume that this correlation between pregnancy and the development of the mammary glands is caused by chemical agents or hormones which pass from the womb into the blood and excite the galactogenic activity of the glands, the phenomenon loses none of its purposive aspect, since it is evident that the hormones in question are here again only acting as

[1] Driesch, *op. cit.*, pp. 177, 178.

mere releasing agents for the galactogenic activity already potentially existing in the cells of the lactiferous glands.

It must further be noted that this co-operation of physico-chemical agents, which mechanists resort to in these and similar cases to account for the activation of a particular function, does not in any way explain why these hormones and enzymes come into action just at the precise moment when their activity is necessary, nor why their activity stimulates just that particular function which is needed at the moment. " More and more, as physiology has become more searching in its study of the functions of the animal," Johnstone writes, " has it sought to explain the metabolic processes by assuming the intervention of enzymes, until the number of these substances has become legion, and much of the original simplicity of the notion of ferment-activity has been lost. But why do not these enzymes, if they are always present in the tissues, always act ? They must be activated, says modern physiology ; that is, the enzyme really exists in the tissues as a ' zymogen,' or a substance which is not, but which may become, an enzyme ; or they exist as ' zymoids,' that is, substances which appear to be chemically enzymes, but which must be activated by ' kinases ' before they can become functional . . . But why do the inert zymoids become activated by the kinases just when they are required by the general economy of the whole organism ? "[1]

All these functional activities which are adapted to their end of maintaining the organism in its normal stationary physiological state, or of bringing it back to that state, or of preparing it in anticipation for certain other functions having similar ends, thus assume an appearance analogous to that of the instincts : " We ought not," writes Johnstone, " to think of diaphragm and lungs apart from the

[1] Johnstone, *op. cit.*, pp. 327–328.

movements of these organs, and we do not say that the first breath drawn by the newly-born mammal is an instinctive action, involving the use of inborn bodily tools —the diaphragm, lungs, etc. We ought not to think of the lips and mouth and pharynx of the young baby apart from the action of suckling the mammae of its mother, but usually we say that this action is an instinctive one. *Where does the ordinary functioning of an organ end and its instinctive functioning begin?* Are the muscular actions of the lobster, when it frees its body and appendages from the carapace during the act of ecdysis, instinctive ones? Most zoologists would say that they are not, any more than the movements of the maxillipedes in respiration are instinctive ones, yet they probably would not hesitate to say that the action of the 'soft' lobster in going into a rock crevice is instinctive."[1]

This great similarity between physiological functional activities, on the one hand, and reflexes and instincts—the psychic or nervous nature and the purposive or teleological character of which are generally recognized—on the other, this passing by insensible degrees from one to the other, argue, therefore, in favour of the nervous nature and purposive character of physiological processes in general, and consequently against the mechanist thesis according to which it would be possible to give an exclusively physico-chemical explanation of these processes.

But though the reflexes and instincts could from certain points of view equally well be dealt with in this chapter on the morphological and physiological phenomena of pre-established adaptation, yet, because of their special nature which brings them into alignment with psychic phenomena properly so-called, it seems preferable to speak of them

[1] Johnstone, *ibid.*, p. 286.

later in a separate chapter, after the examination of the behaviour of the lower organisms.

We will, therefore, confine ourselves here to a résumé of the matter which has been the subject of this chapter and the conclusions which emerge.

1. The morphological phenomena of pre-adaptation to environment have a purposive character, which no object of the natural inorganic world, so long as it has not been touched by the hand of man, possesses.

2. The anticipating co-ordination which exists between the different parts of the organism, even before they have begun to function, and which is in such close relation to their future functioning; the co-ordination which exists between all the means of attack and defence with which organisms are furnished, even before their struggle for existence begins; finally, all the other similar structural co-ordinations which exist in anticipation of future activities—all these co-ordinations have a strongly marked purposive aspect and are peculiar and exclusive manifestations of life.

3. With regard to every element or organ or function of the organism the question: of what use is or has it been? can always be put. Such a question can never nor ever has been asked with regard to any element, body or process of the natural inorganic world, so long as it has remained untouched by man.

4. The Darwinian explanation, which entrusts to a game of chance, in which the organism preserves an aloof and passive attitude, the construction of this same organism (that is, the production of a phenomenon which is the absolute antithesis of phenomena actually produced by chance), proves itself entirely incapable of accounting for even the morphological phenomena of pre-established

adaptation for the explanation of which it was particularly intended.

5. No class of objects in the inorganic world has ever given rise to the need for an explanation of the Darwinian type; we can, therefore, say that it is in itself an implicit recognition of the peculiar nature of the phenomena of life. At the same time it explicitly assumes that life has certain characteristics which in themselves are sufficient to distinguish it from all processes of the natural inorganic world.

6. If we pass from the morphological to the physiological phenomena of pre-established adaptation we see that the tendency of the organism, as revealed by these phenomena, to maintain itself in a stationary physiological state is shared by no natural physico-chemical process. We saw above that the same can be said of the analogous tendency in the elementary metabolic process, of which that of the whole organism is only a direct consequence. This tendency, which is inherent in even the most elementary phenomenon of life, defies any Darwinian explanation. In addition there is the fact that the physiological procedures by which the organism is able to satisfy this tendency are co-ordinated and ready to function even before the need for them arises.

7. The physico-chemical explanations of the production at a given moment of a particular process of self-regulation or functional anticipation are completely unable to explain either the entry into action, always and only at the precise moment at which it is required, of a given chemical agent, or the purposive aspect of these chemically produced processes.

8. The purposive aspect of these processes is so pronounced that the physiological functioning of the different organs merges insensibly into the instincts

CHAPTER IV

PURPOSIVENESS OF THE PHENOMENA OF NEW ADAPTATION

The phenomena of adaptation to new environmental conditions have all a purposive aspect; moreover, they have all the same single 'purpose': the maintenance or the re-establishment of the stationary state of some physiological process. The Darwinian theory of natural selection can offer no explanation at all for this capacity which is inherent in life. In no physico-chemical process of the natural inorganic world is there anything resembling the production of structures or substances for its own 'defence,' 'repair' or 'restoration,' or the production of the appropriate antitoxins for the neutralization of the corresponding toxins. Of no process in the inorganic world can the words 'injurious agent,' 'disease,' 'poison' and still less 'self-healing force,' be used. No concept similar to that of 'function,' 'division of labour,' 'ageing,' or 'death,' can be applied to brute matter. Consequent necessity of recognizing an energetic process *sui generis*, such as we have suggested, to account for the phenomena of new adaptation which are peculiar to life.

WHILE the purposiveness of the morphological and physiological phenomena of pre-established adaptation is self-evident—so much so that even mechanists have recognized it and have been driven to the Darwinian theory of selection in their attempt to find an explanation—the purposiveness of the morphological and physiological phenomena of new adaptation seems to many people less evident. It has even been denied by mechanists attempting to bring these phenomena into the usual categories of the natural inorganic world, in which every physico-chemical process always enters into equilibrium with new energetic environmental conditions as they arise.

But a purposive aspect subsists in these phenomena of new adaptation, for they all have directly or indirectly the single common end which we have so frequently insisted upon: namely, *the stationary state of some physiological process.* They achieve this end either by restoring, with a minimum of accessory modifications—transitory or permanent—the old stationary state which has been disturbed by the new environment or agent; or else, when this restoration is no longer possible, by settling into a new physiological state, *which in its turn is also stationary.* It seems as if the vital process cannot continue unless it succeeds, sooner or later, in one way or another, in acquiring or re-acquiring physiological invariance of some kind.

On the contrary, nothing of this kind takes place in the physico-chemical processes of the natural inorganic world, which, though they continually place themselves in equilibrium with new environmental energetic conditions as they change, never tend *spontaneously*, that is without being prepared for this by man, to place themselves in a stationary energetic state, nor strive to keep themselves in such a state, when disturbing agents intervene, either by escaping from these agents or by neutralizing them, nor adopt a new stationary energetic state when the earlier one cannot be restored in any way. The same observations apply here as were made with reference to the same tendency shown by metabolism, the most elementary and the most general vital phenomenon, to maintain energetic invariance.

An unsuccessful attempt has been made, as we have seen, to explain the purposive aspect of pre-established adaptation by means of the Darwinian theory of natural selection; but a similar attempt to explain in this way the faculty for adaptation to new environmental conditions, and so to account for its purposive aspect, cannot even be dared. We have here, in fact, to explain not such or such particular form of adaptation which some species has gradually assumed in the past, but the faculty of adaptation itself which is inherent in life. And to explain life we can hardly bring in the struggle for life which presupposes life itself.

Thus, in dealing with this faculty of new adaptation with its strongly marked purposive character, the extreme mechanists find themselves driven from their last refuge of Darwinism.

Undeniable evidence of the purposive aspect of this faculty of new adaptation is to be found in the facts in which it is demonstrated. And, be it noted, we are not speaking for the moment of the phenomena of *flight* or *withdrawal,* when the organism strives to *escape* from the new environment or disturbing agent, or in other words of the *movements of animals,* which we shall consider in another chapter; for inasmuch as these tend to *remove* the animal from the disturbing factor or to *re-establish* the old environment, they form a category *sui generis* of phenomena of adaptation to changed environmental conditions, the purposiveness of which is so evident that the mere idea of denying it is inconceivable.

But, though less obvious, the purposiveness of the two classes of new adaptation, in the strict sense of the word, which we are going rapidly to deal with in this chapter, is none the less undeniable. In the first of these classes are included the phenomena of *defence* against the disturbing agent, that is phenomena which seek to *shelter* the vital process from this agent; in the second, are included those by which the organism *neutralizes* this agent or even sweeps it into the vital process itself and thus makes it part of the new stationary state.

To the first category belong, for example, all the defensive substances or structures which every living cell produces to protect its vital process against the action of the disturbing agent.

Thus, under too intense solar light which is injurious

to protoplasm the skin darkens in defence: "Strong light," says Conklin, "and especially light of short wave lengths such as ultra-violet, is very injurious to protoplasma, and when the skin of white persons is exposed to such light the living cells suffer 'sunburn.' But another result of such exposure is that the skin becomes more deeply pigmented or 'tanned,' and this screen of pigment serves to protect the living cells from the injurious rays."[1]

In the same way, the thickening of animal skin is merely a simple means for defending the protoplasm against external agents which tend to consume it or are destructive or harmful in some other way: "Moderate friction and pressure on the skin," continues Conklin, "instead of wearing it thin, leads to the thickening of the epidermis and the formation of callosities by which the deeper lying parts are protected. A similar result follows the application of various chemicals to the skin."[2]

Again, special new structures are formed as *means of resistance* to pressures or tractions which threaten the continuance of the vital process, or as *obstacles* to the production of processes which might disturb the stationary physiological state. Bones which have been wrongly set, for instance, change their structure and develop at the points which will henceforth be subjected to a greater pressure (a pressure which would certainly disturb the vital processes going on in the soft protoplasm and which might finally become fatal for this protoplasm) characteristic thickenings, formed by the depositing of osseous substance and making a solid framework protecting the living substance contained within. Similarly, if the weight of certain fruits is artificially increased, their stalks, which are subjected to a stronger traction, receive addi-

[1] Conklin, *op. cit.*, p. 323.
[2] *Id.*, *ibid.*, p. 323.

tional strengthening. The epidermis of the leaves of a plant transported into a new climate which is too hot and dry becomes thicker and thus more impermeable to evaporation. "A very great number of vegetable or animal tissues," observes Driesch, "possess the property of being determined, as to the degree and the modalities of their development, by the intensity and the specific nature (for instance, by the direction) of their function: the bones and the conjunctive tissues adjust their structures to the direction of the pressure or traction to which they are subjected; in plants the intensity of transpiration influences the development of the epidermis and the conductive vessels; they may also, up to a certain point, prevent themselves from being bent down by regulating in a corresponding manner the development of their mechanical tissues."[1]

All these facts and innumerable others with which everyone is acquainted show how the disturbing action of certain agents is transformed, by the very fact that it is disturbing, into a genuine trophic action for the substances and structures whose function it is to defend particular parts of the organism against such disturbing agents.

Nothing comparable occurs in the physico-chemical processes of the natural inorganic world. They never produce substances for ensuring their own defence—that is, "for the purpose of" keeping them in their unvarying energetic activity. Very differently from the living substance, physico-chemical processes, when exposed to new environmental conditions, simply transform themselves, and as it were unresistingly, into new forms of energy or into new variants of their present form of energy, so as to re-establish equilibrium between them and the changed conditions of the environment.

[1] Driesch, *op. cit.*, p. 179.

"We can characterize protoplasm," Lillie writes, "as a *regulated chemical system*, of such a kind that *disturbance* of the normal structure or composition at once determines *constructive or reparative processes* that tend to restore the normal condition."[1]

No physico-chemical process in the natural inorganic world appears to be regulated in this fashion; and further, the very word 'disturbance' is without meaning if it is applied to any of these processes; not one of their conditions can be described as being 'normal' or 'abnormal'; and they do not offer a single example of such reparation or restoration of a normal state.

From these cases of *defence*, whether external or internal, against disturbing agents, in which the adaptation consists for the most part in protecting the vital process from the disturbing agent, we pass to what can be called the physiological processes of *attack* against the disturbing agent itself, processes whose 'purpose,' that is to say, effect, is to neutralize it or to render it incapable in some way or other of continuing to be injurious.

The classical experiments of Dallinger and Davenport, for instance, on the acclimatization of the lower organisms have proved that infusoria can be gradually accustomed to endure successively higher temperatures, so that they can live in temperatures which would kill a similar organism which had not been acclimatized. Similarly it is possible gradually to accustom fresh water amœbæ and infusoria to live in water of a degree of salinity which would at first have destroyed them. And it seems as if this adaptation is caused by the production of new sub-

[1] R. S. Lillie, "The Transmission of Physiological Influence in Nerve and other Forms of Living Matter," *Scientia*, December, 1920, p. 435.

stances which are able to absorb the excess of heat or to neutralize the action of the salts.[1]

Adaptation to certain mineral substances which are poisons for the normal organism is no less typical. "Even more remarkable," writes Conklin, "are the adaptations that organisms show to certain poisons, if these poisons are given in graded doses so that the organism acquires a tolerance for them. Such tolerance may be acquired to a limited extent for violent mineral poisons, such as corrosive sublimate, as Davenport showed in the case of Paramecium. It is also known that human beings, as well as other organisms, may acquire tolerance for arsenic and arsenical compounds."[2]

This adaptation is still more characteristic in cases where it can be shown to be due to the fabrication of anti-bodies or antitoxins by the cell which are designed to neutralize the action of the poison: "It is certain," continues Conklin, "that the ancestors of the guinea-pig, which is a native of South America, could never have had any experience with the venom of the cobra, a native of India; and yet the guinea-pig can form an anti-body against cobra venom, and for every particular kind of venom its own particular anti-body."—"Each kind of toxin leads thus to the formation of a specific anti-body, which serves as an antidote for that poison. Many of these toxins are complex and highly specific substances, and yet the living organism, if given sufficient time, can make a specific antidote for each particular kind of toxin."[3]

Even Bohn, vigorous anti-vitalist though he is, writes:

[1] C. B. Davenport and W. E. Castle, "On the Acclimatization of Organisms to High Temperatures," *Arch. f. Entw. Mech. der Organismen*, Vol. II, Part II, July, 1895; C. B. Davenport and K. V. Neal. "On the Acclimatization of Organisms to Poisonous Chemical Substances," *ibid.*, Vol. II, Part IV, January, 1896.

[2] Conklin, *ibid.*, p. 324.

[3] *Id.*, *ibid.*, p. 325.

"The organism *defends itself* against every invasion of foreign bodies which may disturb *the harmony of exchanges* by producing ferments suited to each particular case." —"For every toxin which is introduced into the organism the latter retaliates with a corresponding and specific antitoxin."—"The introduction of an alien substance into an organism causes the more or less rapid formation of an opposing substance to counteract its effects. The presence of a 'poison' leads the organism to respond by producing a 'counter-poison.'"[1]

It is evident that the formation of pigments as a means of defence against solar rays, the production of callous substances as a means of defence against the mechanical action of waste, the production of anti-bodies as a means of defence against poisons—in short the formation of new substances which are so many means of organic adaptation to changes in the environment—have one and the same end, viz., to protect the organism against the disturbing agent or to neutralize this agent or to involve it in the vital process, and so to allow the organism, or any one of its parts and particularly those which are most directly concerned, to re-enter a stationary physiological state, either the one which existed before or a new one.

Tanzi and Lugaro, who are both extreme mechanists, write: "To react means to adapt ourselves, either momentarily or permanently, to new factors, sometimes ephemeral, sometimes lasting, which are modifying the conditions in which we live. Every reaction is, therefore, *simply a change which is necessary and advantageous for the continuance of life*."[2] It would seem that they do not realize that in speaking in this way they are endowing life with a

[1] G. Bohn and A. Drzwina, *op. cit.*, pp. 69, 213, 246–247.

[2] E. Tanzi and E. Lugaro, *Malattie mentali*, 2nd edn., Milan, 1914, Vol. I, p. 216.

very special character and are clearly recognizing that in each of these reactions it is pursuing the single and unique 'purpose' of *its own conservation*. It is this characteristic which distinguishes it from the whole of the natural inorganic world.

Whence does the living substance derive this marvellous property of persistence, this capacity for defensive reaction against all that threatens or disturbs it, a property which, we repeat, so clearly differentiates it from brute matter ?

Here again it is evident, as we have already seen in the previous chapter, that the tendency of the living organism as a whole to remain in or enter a stationary physiological condition—preferably the one in which it was before it was disturbed, or, if that is impossible, some other—follows directly from the same tendency which is seen in every single cellular metabolism, the most elementary and general vital process. Living protoplasm, disturbed in its normal metabolism *has no rest* until it succeeds in re-establishing the old or in producing a new stationary metabolic process ; that is : as an inevitable consequence of losing its dynamic 'stationarity' the metabolic process passes rapidly, almost violently, and without ceasing, through successive assimilative and dissimilative phases, all different from each other, with a corresponding chaotic production of different substances, till among these it fortuitously produces one capable either of neutralizing the disturbing agent or of bringing about a new stationary metabolism, which includes among its elements this very agent, which thus ceases to be a disturber. With this return to a stationary metabolic state, whether it is the old one or a new one, that is with the re-establishment of an unvarying process, the continuous production of new substances ceases *ipso facto*, and all that continues is the production of the substance which succeeded in bringing the organism back to a stationary state.

We shall find it profitable to examine this point more closely. We saw in considering metabolism that the substance which is destroyed in each of the dissimilative phases of the metabolic process tends to reconstruct itself, both qualitatively and quantitatively, during the immediately succeeding assimilative phases. It is the excitation or the nervous discharge itself, which is produced in the nucleus at the moment and as a consequence of the explosive process of dissimilation, which tends later, during the second phase (or assimilative phase) of the oscillating intra-nuclear discharge of nervous energy, to reproduce, in accordance with our hypothesis of specific accumulation, the very substance, identical in quality and quantity, whose decomposition previously gave rise to this same excitation or nervous discharge. This tendency to reproduce exactly, in a second phase, the substance which has been destroyed during the preceding phase, constitutes the tendency of the metabolic process to remain in a stationary state, as a number of writers, and particularly Hering, have pointed out.[1]

Now let us suppose that some disturbing factor interferes in such a stable metabolic process: a poison, for instance, by decomposing or transforming or attracting to itself some of the indispensable atomic groups in the nutritive liquid, prevents the substance which has just been destroyed from being reconstituted as it would have been if the metabolic process had not been disturbed. The succeeding phases of dissimilation and assimilation will still continue, but each assimilative phase will no longer reproduce the identical substance which has been

[1] *Cf., e.g.*, E. Hering, *Zur Theorie der Vorgänge in der lebendigen Substanz*, pp. 54–55.

decomposed during the previous dissimilative phase, since the components of this substance, having been decomposed or transformed or attracted by the interfering poison, will no longer be found in the nutritive fluid. The result is that each succeeding dissimilative phase will differ in quality and in quantity from the immediately preceding one; hence there arises a series of dissimilative phases all different from one another, which will give rise to the disordered production of incessantly new materials. But as soon as there appears among these substances one which will enable, either directly or indirectly, a stationary metabolic state to be re-established, one, that is to say, which creates conditions such that the assimilative phase following immediately upon the dissimilative phase which has produced this substance becomes once more identical, in the contrary direction, with this last dissimilative phase, then this stationary metabolic state will straightway be restored. This leads *ipso facto* to a halt in the production of those dissimilative phases which were all different from one another and which gave rise to the incessant formation of new and varying materials. To their place will succeed identical dissimilative phases which will continue to reproduce this substance—the anti-body which neutralized the disturbing effect of the poison—as that which fortuitously succeeded in bringing the metabolic process back to a new stationary state.

It thus ceases to be *a mystery* " why a particular toxin always causes the formation of its one appropriate antitoxin."[1]

But this mystery is dispelled only if it is recognized that the metabolic process tends in all cases to settle into a stationary state. And it cannot too often be repeated

[1] Conklin, *op. cit.*, p. 339.

that no physico-chemical process in the natural inorganic world shows this tendency.

Immunization, which is due to the production of antitoxin in excess of the quantities strictly necessary, becomes equally comprehensible: "When the organism reacts against microbes which are poisoning it by the production of antidotes," writes Grassi, "it does not confine itself to producing only the quantity which is strictly necessary, but goes beyond that limit and forms an excess, and man can make use of this excess in preparing the sera which he employs against microbes."[1]

Driesch in his turn writes: "In the formation of antitoxins the organism possesses a faculty which is primarily important as a means of regulation: it reacts to the majority of animal and vegetable poisons by producing, in excess, a counter-poison. Immunization is based on this fact."[2]

Now this overproduction of antitoxin, in which the immunization of the organism even for the future and the possibility of the curative application of sera to other organisms have their origin, is the direct consequence of the fact that after the chaotic efforts of the disturbed metabolic process have culminated in the fortuitous production of the antitoxin which is able to re-establish a stationary state, the production of this antitoxin continues during all the succeeding dissimilative phases which are now all identical again.

The famous *vis medicatrix naturæ*, whose pronounced purposive and providential character has aroused the admiration of so many biologists, springs thus from this very tendency of the metabolic process to free itself from or neutralize the effects of any foreign and unaccustomed

[1] G. B. Grassi, *La Vita. Ciò che sembra a un biologo*, p. 5.
[2] Driesch, *op. cit.*, p. 178.

agent, especially if it be of a pathogenic nature, which is disturbing it, and so to re-enter the old or a new stationary state. As Grassi writes: "If pathogenic microbes invade our organism it seeks by its own efforts and without our knowledge to produce substances which either destroy the poison secreted by the microbes, or kill the germs, or prevent the production of them, or finally destroy their virulence without injuring their vitality. Thus the organism acts at the same time as its own doctor and its own very skilful chemist."[1]

Nothing similar to this ever occurs in any physico-chemical process of the natural inorganic world, where concepts such as 'injurious agent,' 'disease,' or 'healing force' do not exist and would be meaningless; for all these concepts, and especially the last, imply *an activity of resistance* to disturbing agents, which is absolutely wanting in non-living matter.

To recapitulate, we can say that following upon the introduction of a disturbing element to which the organism has "to adapt itself," there comes a phase of metabolic instability of indefinite duration and characterized by an "over-production of reactions," and of these the only one which *persists* and becomes the *suited reaction* is that which has fortuitously succeeded in bringing the metabolic process to a new stationary state.

This process has been compared, and with good reasons from certain points of view, by Conklin and Zur Strassen (by the latter under the name of "over-production of opportunities"), to the over-production of movements and to the process of "trial and error," by which the majority of behaviourists, and especially Jennings, have sought to explain those adapted movements of both higher and lower organisms which we shall examine in the next chapter.

[1] G. B. Grassi, *ibid.*, p. 6.

In the case we have been considering, it is a question of a physiological process of trial and error on the part of metabolism which tries by all means to escape from its state of instability into one of energetic invariance.

In other words, the tendency inherent in every vital process, whether simple or complex, to remain in or enter a stationary energetic state is expressed in the 'selection'—that is, in the transformation of the fortuitous into the permanent—of those conditions, external and internal, which make possible the persistence or restoration of physiological stability.

This explanation will equally account for other peculiarities of the metabolic process of new adaptation. The feverishness, for example, which generally accompanies metabolic instability throughout its undefined duration, can be explained, if our hypothesis is accepted, as the result of the fact that the heat-producing, disintegrating processes of each dissimilating phase predominate over the reintegrating processes, characterized by the absorption of heat, of the succeeding assimilative phase. Hence the sick person experiences a rise in temperature which persists until the restoration of qualitative and quantitative identity between the disintegrating and reintegrating processes causes the residual excess of heat to disappear: "It is known," says Conklin, "that immunity to bacterial or other toxins is not acquired immediately but only after a certain lapse of time, during which physiological processes are more or less disturbed; there is frequently an increase of destructive metabolism, the body temperature rises, and there are other abnormal conditions. Fever, say Adami and McCrea, is the process of adaptation to such toxic agencies as can be neutralized by the development of anti-bodies."[1]

[1] Conklin, *op. cit.*, p. 346.

The explanation which we have given above of the inherent nature of the process of adaptation also accounts for the specialization of functions and the division of labour which have gradually been produced in the different parts of the organism in the course of phylogenetic evolution. A disturbing agent which is capable of affecting the whole organism, for example, a toxic antigen, or some waste product of certain tissues which exerts a toxic or disturbing action on the other tissues, circulates in the blood and thus exerts an injurious influence on all the cells of the organism or on a great part of them—throws the cells of the organism into a state of 'uneasiness' and 'restlessness,' which is evinced in the incessant and disordered production of new substances by the metabolic process which has lost its stationary condition. Now the group of cells which by chance first succeeds in producing the required anti-body, either for neutralizing the disturbing agent or for incorporating and absorbing it in its metabolic process, thus succeeds in suppressing the disturbance, not only on its own account, but also for all the other cells of the organism. By remaining henceforth in this newly acquired stationary state, whilst all the other cells have returned to their old stationary state, the cellular group in question thus assumes permanently this new and useful function, with its corresponding specialization, in relation to the rest of the organism.

We can thus understand, for example, how the function of eliminating from the blood certain waste products which would be harmful to the whole organism came to be assumed by the kidney, and again how the production of various hormonic substances useful for the whole organism or for one of its parts has been taken over by the various internal secretion glands.

At the same time we can see how this specialization of

functions assumed by the various organs, tissues and cells, fixing and consolidating itself more and more as a result of the continual increase of specific mnemonic accumulations in the nuclei, gradually diminishes the faculty of new adaptation possessed by young cellular groups; for in new adaptation it must be possible for new and different specific mnemonic accumulations to be deposited—without hindrance from the strong activity of the others or from deficiency of space or nutrition—in addition to those which are already in existence. The diminution of the capacity for adaptation, thus caused by the constant growth of mnemonic accumulations relative to the various specializations of functions, produces " old age " in these organs, tissues and cells and, consequently, in the whole organism formed by them. As a result of this " old age," and particularly when it has become very pronounced, the organism ends by losing its aptitude for adaptation to changes in its environment, even when they are of the slightest, so that eventually the merest trifle will cause its death. Minot in particular endeavoured to establish the fact that the death of metazoa is due to the greater differentiation and specialization of their tissues. " Admitting the immortality of the unicellular organisms, he argues that death is the price metazoa pay for the higher differentiation of their cells."[1]

In conclusion let us summarize the facts and considerations brought forward in this chapter.

1. The phenomena of adaptation to new environmental conditions possess a common purposive aspect; further, they all have one and the same end: viz., the maintenance or the re-establishment of a stationary state for the

[1] Loeb, *op. cit.*, p. 362.

particular or general physiological process affected by the change. Nothing in the natural inorganic world possesses anything remotely resembling this tendency, so long as it is free from man's interference.

2. The Darwinian theory of natural selection which has been put forward, though unsuccessfully, as an explanation of the purposive aspect of the phenomena of pre-established adaptation, cannot even begin to explain the faculty of new adaptation which is inherent in life.

3. No physico-chemical process in the natural inorganic world makes any approach to the production of substances or structures for the purpose of its own 'defence,' 'reparation,' 'restoration,' or 'immunization.'

4. Above all in non-living matter we see nothing resembling the process by which the presence of a particular toxin causes the production of the corresponding antitoxin, and of only that particular antitoxin which is able to neutralize the toxin or to destroy or render harmless the different bacteria.

5. There is no process in the natural inorganic world to which we can apply the words 'injurious agent,' 'poison,' 'disease,' or, even less, the phrase a 'self-healing force' inherent in the process itself. When applied to the natural inorganic world these concepts are completely meaningless.

6. Feverishness, the phenomenon of reaction against harmful processes, is equally unknown outside life.

7. Nothing similar to the concepts of 'function' or 'division and specialization of labour,' or to the processes of 'ageing' and 'death' is seen in brute matter which is not animated by the breath, by the rhythmic pulse of life.

With these facts in mind, we must again ask whether, instead of closing our eyes before the most obvious reality and obstinately repeating that the properties of dead matter are sufficient to account for those of life, is it not more in keeping with a positive method to try to determine what vital phenomena have in common and what differentiates them from inorganic matter; in other words, is it not more in keeping with positive method to try to explain by some hypothesis upon the nature of the vital process the properties common to these phenomena by which they are differentiated from all inorganic phenomena?

The explanation which we have suggested for the facts rapidly surveyed in this chapter seems to show that it is possible to venture upon such a hypothesis; and although we cannot certainly say that it fully explains them all, it nevertheless indicates that the path we have taken, namely, the search for a subsidiary hypothesis to explain life, is the right one and the only one which allows us to hope for fruitful results.

CHAPTER V

PURPOSIVENESS OF THE BEHAVIOUR OF THE LOWER ORGANISMS

Movement in the lower organisms is shown by its fundamental characteristics to be the result of forces inherent in the latter. Their behaviour is always directed towards one and the same end: the preservation of their optimum environment. The concept of an optimum environment becomes meaningless when applied to purely physico-chemical processes. Nothing remotely resembling 'avoiding reactions' is ever seen in any physico-chemical process of the natural inorganic world. Nothing remotely resembling 'trial and error' behaviour is ever seen in brute matter. Nothing remotely resembling 'learning' by experience is ever seen in any particle of non-living matter. Childish simplicity and complete inadequacy of the theory of tropisms. We must once again, therefore, resort to biological memory, a property pertaining exclusively to life, in order to account for the purposive behaviour of the lower organisms.

IN this and succeeding chapters we can proceed more rapidly since the purposiveness of the phenomena which we have still to consider becomes more accentuated as we pass from the most elementary and strictly physiological vital phenomena to psychical phenomena in the narrow sense.

The behaviour of even the lowest organisms is clearly differentiated from that of brute matter; the animalcule moves by its own inward forces, it is autonomous in its movements, and so long as it is alive it never abandons itself passively to the play of external energies, like a piece of wood left to the mercy of the waves.

Sufficient proof that the behaviour of these lowest organisms has a distinct and undeniable purposive aspect is afforded by the simple observation that its sole aim is to keep the animal in its optimum environment or restore it to it when change intervenes.

Thus, for example, when the infusorian *Paramecium* is at a temperature of 28 degrees it reacts negatively to an increase but not to a diminution of temperature; when, on the other hand, it is at a temperature of 22 degrees it reacts negatively to a diminution but not to an increase of temperature. And, again, *Euglena*, when in a moderately lighted environment, reacts negatively to a decrease but not to an increase of illumination; while the opposite happens if it is exposed to a strong light.[1]

Oysters and actiniæ when exposed to the air close up—behave, that is, in such a way as to keep the humidity of their environment unvaried.[2]

Environmental invariance also includes the position of the organism in relation to the various external forces to whose influence it is subjected, and, in particular, to the force of gravity; hence the tendency to preserve or re-establish the animal's accustomed support or its normal position. Thus, for example, the amœba generally draws in its pseudopodia when it comes into contact with inedible solid bodies; but if it is taken up from the bottom of the aquarium and suspended in the middle of the water, it extends its pseudopodia in all directions and as soon as one of them touches some solid body it attaches itself to it, brings its whole body there and rests upon this new support in its accustomed manner. The starfish when turned over tries to right itself, that is to return to its normal environmental conditions in relation to the force of gravity. The actinia *Cerianthus* behaves in a similar way; if it is placed head downwards in a vertical tube of sea water, it, too, in spite of the difficulties caused by lack of space, begins to try to right itself and will often spend more than an hour before it succeeds.

[1] H. S. Jennings, *Behavior of the Lower Organisms*, New York, 1906, pp. 294, 295.

[2] H. Piéron, *L'évolution de la mémoire*, Paris, 1910, pp. 29, 74.

Examples of a similar nature could be enumerated indefinitely.

The mere fact that organisms have an optimum environment substantially differentiates them from any physico-chemical process of the inorganic world. The expression is without meaning when applied to the latter. An iceberg, for instance, moving towards the warm seas of the south, shows no signs of anything approaching an 'avoiding reaction' by which it displays its preference for remaining in cold water.

And yet Loeb writes: "From a physico-chemical viewpoint . . . environment cannot affect the living organism and non-living matter in essentially different ways."[1]

But if the effect of environment on the living organism is no different from its effect on non-living matter, how does it come about that for all living organisms without exception, and only for these, there is an optimum environment which they do all in their power to preserve?

The attempt made by organisms, even the lowest, to avoid any change which would remove them from their optimum—and the optimum environment is only that which is habitual—is, evidently, simply the consequence of their tendency, to which we have frequently drawn attention, to maintain a physiological stationary state, from which tendency, as we have seen, their capacity for adaptation is also derived. Instead of reacting to the disturbing agent with the metabolic production of some counter-agent or anti-body which might neutralize or destroy it, the organism reacts in this case by avoiding it. Nuclear discharges are produced which impart to the fibrils of the protoplasm certain contractions and distentions; these produce certain movements in the animal and remove it from the disturbing agent itself or enable

[1] Loeb, *op. cit*, p. 286.

it in some way to avoid the disturbance which the agent causes in its environment.

That the two processes, the metabolic production of anti-bodies and the production of movements of avoidance, both resulting from modifications of the normal nuclear activity of the metabolic process, are fundamentally of the same nature is proved by the analogy (noticed, as we have seen, by other writers) between the phenomena with which these two processes begin. In the one case there is an over-production of substances and in the other case an over-production of movements, followed in both cases by the 'selection' of the reaction which has been fortuitously successful. that is by the transformation of a transitory into a permanent reaction.

Jennings in particular has observed and described the 'avoiding reactions' and 'trial and error' behaviour of unicellular and other lower organisms in a keen and penetrating manner unclouded by mechanistic prejudices.[1]

"In this unsystematic groping," says Claparède when considering the question of 'trial and error,' "the selection of the favourable reaction is mechanically determined by external circumstances: the surrounding medium sorts out the reactions and only allows those which are able to solve the situation to continue."[2] By this he means that it is the environment which selects and fixes the favourable reaction, in the sense that when the chance production of the latter has restored the environment to its previous state, then the environment, allowing the return to the normal physiological state, by that very fact causes the chaotic production of further movements to cease.

This system of 'trial and error,' however, only comes

[1] H. S. Jennings, *op. cit.*, passim.

[2] E. Claparède, "La psychologie de l'intelligence," *Scientia*, November, 1917, p. 359.

into play when *totally new* disturbing circumstances arise. When the same disturbance has been repeated several times, mnemonic association, as a result of the mnemonic property of living substance, is established between the initial impression or 'sensation' produced on the organism by the disturbing agent at the moment of its introduction and those movements which have enabled it in the past to avoid this agent. Hence these movements can be produced immediately, without any tentative efforts, as soon as the disturbing agent appears.

The animal thus 'learns' to adapt its movements to changing external circumstances.

It is now definitely proved that this mnemonic learning takes place even in lower organisms which are devoid of a nervous system. And we must not forget that by this learning the first awakening of what in the higher organisms we call 'intelligence' is brought about. "The protozoan *Paramœcium*," writes Johnstone, "studied by Jennings, or the crabs, crayfishes and starfishes, studied by Yerkes and others, really *learn* to perform actions, but this learning is said to be the result of 'trial and error.' The animal tries one series of movements and finds that it fails, tries another and another with a similar result, and in the end finds one that is effective. This is remembered, and when the same problem again confronts the animal it is solved after fewer trials, and finally, after experience, the end-result is attained at once without previous trials. Now many of what we call truly intellectual processes are certainly processes of precisely this nature."[1]

The behaviour of the infusorian *Stentor*, which Jennings so ably describes, is typical in this respect. When he disturbed this infusorian, which was attached by its

[1] Johnstone, *op. cit.*, p. 293.

peduncle to the bottom of the aquarium, with a constant flow of carmine, he first observed that the animal bent to one side, as if to avoid the disturbance. This 'avoiding reaction' was repeated several times, but when it did not succeed in this way in protecting itself from the dense cloud of carmine grains, the animal changed its reaction. The ciliary movement was reversed in direction, so that the particles against the disk and in the pouch were thrown off. This lasted only for a moment, and then the animal returned to its normal ciliary movement in the same direction as before. But when after this the grains of carmine continued to molest it, the reversal was repeated two or three times in rapid succession. Failure to escape from the disturbance being once again the result, the tiny organism adopted a new reaction and contracted into its tube of mucus. This enabled it to achieve its end, but at the cost of suspending its activity and losing all opportunity of obtaining food. Hence it remained in the tube not longer than half a minute, and then emerged and resumed its ordinary ciliary movement. But when it once more found itself disturbed by the grains of carmine *the animalcule, instead of repeating the first two attempts which had proved to be useless*, namely, bending to one side and reversing the current, *immediately and without hesitation resorted to the third*, that is, to withdrawal into its tube. And, finally, when repeated recourse to this third reaction proved unavailing the animalcule ceased to extend itself, freed itself from the bottom by means of vigorous jerks, and swam away to find a new resting place.[1]

When faced with these typical examples of purposive behaviour in the lower organisms—and of a behaviour which profits by experience—it would have been thought that extreme mechanists would, at least in cases of this

[1] H. S. Jennings, *Behaviour of the Lower Organisms*, pp. 174, 175.

category, have admitted that life appears to be endowed with special fundamental properties not possessed by any physico-chemical process of the inorganic world. None of these reveals, in fact, anything similar either to the tendency to maintain invariable environmental conditions, or to the mechanism of reactions by which the organism secures invariance in relation to its environment, or to the process of learning as a result of which present and future behaviour depends on the successes and failures of the past.

Nevertheless certain mechanists, and these not the least, as the name of Loeb shows, have attempted in their theory of tropisms to give a mechanistic explanation of the behaviour of the lower organisms. But they have only succeeded in giving yet another illustration of the grave dangers which preconceived ideas involve and which may even lead to an entirely false presentation of observed facts.

The way in which Loeb attempts to explain the behaviour of certain of the lower organisms by his theory of tropisms is well known: "If a positively heliotropic animal is struck by light from one side, the effect on tension or energy production of muscles connected with this eye will be such that an automatic turning of the head and the whole animal towards the source of light takes place; as soon as both eyes are illuminated equally the photochemical reaction velocity will be the same in both eyes, the symmetrical muscles of the body will work equally and the animal will continue to move in this direction."[1]

This explanation ought to be equally valid for the attraction which artificial light has for certain winged insects. But, unfortunately for Loeb, merely an evening walk in any lighted street will supply enough evidence to

[1] Loeb, *op. cit.*, p. 259.

prove that it is quite untrue that insects fly directly towards the light of the lamp. They approach it obliquely, describe a number of circles round it and then, if their wings have not been singed, fly away from it. Where in this case is the directness of movement towards the source of light which we ought to see if Loeb's theory is correct ?

And there are besides innumerable other facts to show how artificial and childishly simple is the explanation of animal behaviour which Loeb supports—all the 'avoiding reactions' by which infusoria and other lower organisms endeavour to maintain themselves in their optimum environment ; the behaviour of 'trial and error' of the lower organisms, which Jennings has described so well ; the behaviour, which we have described above, of the infusorian *Stentor* when stimulated by grains of carmine ; the contorted movements and the attempts to escape, which Jennings has described, made by the small amœba pursued by a larger amœba which finally swallowed it ; the varied struggles which take place and are so furiously pursued even among multicellular organisms in the lowest degrees of the zoological scale, and which are so rich in unexpected happenings and acts which are as far removed as possible from mechanical precision (as, for instance, the disarming of a sea-urchin by a starfish) ; the whole behaviour of the higher organisms to which Loeb would presumably not wish to ascribe an essentially different nature from that of the behaviour of the lower organisms ; and finally, the affective tendencies and the general conduct of man which we shall examine later (and we presume that Loeb would not venture to deny that man also is an animal). All these facts argue against Loeb's theory of tropisms which, in substance, places every animal on the same level as the mechanical mouse which children play with, which, when it is wound up and placed between two rails, runs along continually banging itself against one side or the other. In Loeb's explanation the two rails are represented by the parallel solar rays which, by means of the intermediary apparatus formed by the nerves and muscles, keep the unfortunate animal, compelled by its internal mechanism always to go blindly and aimlessly forward, always in the direction of the source of light.

Loeb even tries to reduce the nuptial flight of bees to the mechanism of a mechanical toy! "Kellogg has shown," he writes, "that the nuptial flight of the bees is also a purely heliotropic phenomenon. When a part of the hive remote from the entrance is illuminated the bees rush to the light and can thus be prevented from swarming. These phenomena suggest that the presence of some substance secreted by the sex glands may cause the intensification of the heliotropism which leads to the nuptial flight."[1]

But if the nuptial flight is reduced to such a mechanical process, all the complex and varied modes of behaviour which accompany or succeed it cease to be comprehensible. The bees ought to go on flying in the direction of the sun until they are completely exhausted, in the same way as the clock-work mouse runs along between the rails until its works have run down. This mechanical mouse can be made to go forward when it is wound up and kept in the proper direction by the rails, but we cannot expect it suddenly to change from a purely mechanical toy into a living being and perform actions and display a purposive behaviour which are incapable of being explained in any way by its mechanism.

The childishness of Loeb's mechanical explanation becomes apparent as soon as behaviour begins to be a

[1] Loeb, *op. cit.*, pp. 279–280.

little more complicated. And when, to strengthen it, it is supplemented by the hypothesis that at the required moment the tropisms stop and are turned in the opposite direction, we have not only a further display of childish simplicity but also an implied retraction of the theory first put forward. The validity of our criticisms will be apparent if the following passage taken from Loeb is carefully read and considered : " It is the positive heliotropism of the caterpillars of *Porthesia chrysorrhœa* which makes them move upward where they find their food, when the mild air of the spring calls them forth from their nest. At the top of the branch, they come in contact with a leaf, and chemical or tactile influences set the mandibles of the young caterpillar into activity. . . The animal after having eaten (and having eaten all the leaves at the top of the branch), is no longer a slave of the light, but can creep in any direction. . . The restlessness which accompanies the condition of hunger makes the animal creep downwards—which is the only direction open to it—where it finds new leaves on which it can feed. The 'wonderful hereditary instinct' upon which the life of the animal depends, is thus only its positive heliotropism in the unfed condition and its loss of the heliotropism after having eaten."[1]

First, heliotropic activity which ceases (not, it must be noted, when the animal is completely satisfied, but when there are no more leaves at the top of the branch and it is still hungry), with clock-like precision, so that the animal is always near food when it needs it ; then, the fact that when in contact with its food it always finds in it chemical or tactile influences capable of setting its mandibles in motion and of imparting to them just those movements which enable the food to be introduced into

[1] Loeb, *ibid.*, pp. 280–282.

the digestive canal; finally, the fact that all this is the outcome of a fortuitous combination which quite by chance results in keeping the animal alive—these are the simple facts which to Loeb's mind call for no further explanation!

And if all animal behaviour can be explained by the mechanical theory of tropisms, what is the attraction which the mild air of spring has for the larvæ and which makes them come out of their nests? What is the 'restlessness' which accompanies the state of hunger? What is the new agent, since in this case it is not a question of tropisms, which causes the animal to move downwards as soon as it has eaten all the leaves at the top of the branch? In short what is it which, at a given moment, inspires the mechanical toy with the breath of life?

Loeb and the other extreme mechanists go wrong in that they fail to see that here again in this case of so-called tropisms given external stimuli *do not by their sole action directly and completely determine* these movements of the organism and all their forms; they merely *release*, as a result of mnemonic associations, tendencies which already potentially exist in the organism, and, consequently, the corresponding movements to which these tendencies give rise; and these movements are reproduced in the exact form in which they have been fixed in the past as a result of a longer or shorter period of 'trial and error,' which, in its turn, was stimulated by these same tendencies.

Confusion between *the simple action of release*, performed by given external stimuli, of given tendencies and their corresponding movements and *the directly determining action* is carried so far by Loeb that he tries to mechanize and reduce to a matter of tropisms all animal behaviour —even the most complex, the most varied and the most unforeseen. He sees, for example, in the acts by which the male pursues, courts and wins the female only " combinations of chemotropism and stereotropism."[1]

It is almost impossible to conceive of anything more narrow-minded and stubborn than the attitude of ultra-mechanists such as Loeb, who close their eyes to reality in order that they may not have to abandon theories which are contradicted by facts and which are obviously false !

Without pursuing the question further we will bring this chapter to an end with a summary of the facts and considerations we have brought forward.

1. The fundamental characteristics of the movements of lower organisms show that these creatures move by their own inward forces, that they are autonomous in their movements, that they never abandon themselves passively, as brute matter does, to the mere play of external energies, but react actively to them.

2. The behaviour of the lower organisms in all its aspects has always the same 'aim': the maintenance of the animal in its optimum environment. No object of the natural inorganic world, unless it has been fashioned by man, gives evidence of anything resembling this tendency.

3. The very concept of an optimum environment is meaningless in so far as physico-chemical processes are concerned. They merely enter into equilibrium with changing external conditions, without showing any preference for any particular state.

4. No physico-chemical process of the natural inorganic world shows anything remotely similar to the 'avoiding reactions' of the lower organisms.

[1] Loeb, *ibid.*, p. 284.

5. Brute matter shows nothing remotely resembling the behaviour of 'trial and error' which is typical of these organisms.

6. In no fragment of non-living substance do we see anything remotely resembling behaviour influenced by past experience.

7. Behaviour which is influenced by past experience represents the dawn of what, in the higher organisms, we call *intelligence*. But on the subject of intelligence the most pronounced mechanists are the first to declare (in order to try to retain for their false theories at least facts of a purely physiological order) that it lies completely outside the sphere of phenomena which admit of a physico-chemical explanation.

8. The theories of Loeb and others who try to explain the behaviour of the lower organisms by comparing it with that of mechanical toys are so completely and obviously at variance with reality that they only serve to show the complete blindness which may fall upon even the most distinguished observers and experimentalists when they persist in preconceived ideas and theories which are entirely contrary to established facts.

CHAPTER VI

PURPOSIVENESS OF THE REFLEXES AND INSTINCTS

The purposive aspect of reflexes is so pronounced that they, more than any other manifestation of life, have created the impression that the organism is a 'machine.' The mechanists, however, forget that every machine is constructed for a given purpose and that it presupposes a mechanician who has designed and constructed it. We must, therefore, either admit the 'clockmaker' of Voltaire or have recourse to some fundamental property of life which renders this admission unnecessary. The purposive aspect of the instincts is, if possible, even more pronounced than that of the reflexes. They are of the same nature as intelligent acts, already instructed by past experience. They are, therefore, of an essentially mnemonic origin and nature. Consequently, every attempt to explain them by physicochemical laws alone is condemned to complete failure.

THE purposive aspect of reflexes, especially in the higher organisms, could hardly be more pronounced than it is. Johnstone writes: "Reflexes are, in a way, commonly occurring, purposeful and useful actions, and their object is the maintenance of a normal condition of bodily functioning."[1]

This aspect is so pronounced that the reflexes, perhaps more than any other manifestation of life, have given biologists the impression that the organism is a 'machine' in which all the parts, from the greatest to the smallest, are perfectly adjusted so that they work together in complete co-ordination to perform the functions assigned to the whole.

But in their somewhat too hasty enthusiasm for this comparison between the play of reflexes and that of the parts of a machine the mechanists have forgotten that every machine is constructed *for a given end*, and that it

[1] Johnstone, *op. cit.*, p. 6.

presupposes a mechanician who has designed and constructed it.

Our theory, however, requires no such mechanician, because of the fundamental elementary properties which it assumes as belonging exclusively to life. In fact, it not only takes over the 'trial and error' theory of Jennings and the American school in general to explain animal behaviour, but it also penetrates further into the inner nature of this behaviour by attributing to the metabolic process, and consequently to the whole organism, a tendency to preserve its own stationary state. Without this tendency 'trial and error' behaviour would itself remain unexplained, since 'to try' implies an end to be attained. The theory is completed by the mnemonic property attributed to all living matter, which fixes in the respective specific accumulation the physiological activity constituting the trial which has been successful and therefore 'selected,' and which, by repeated reproduction, finally becomes mechanized in the form of a reflex. In this way mechanician, designer and constructor of the organism-machine are dispensed with.

There is no escaping the dilemma: if we hold that the organism is a machine in which only physico-chemical forces are active, we must then grant Voltaire's 'clock-maker,' who has designed and constructed this delicate and perfect mechanism; or, if we do not wish to grant this creator, we must then have recourse to some fundamental property of living substance which can take his place.

The mechanists make another mistake in that they start from the reflexes, as a basic elementary phenomenon, and then, by compounding them, proceed to an explanation of non-mechanized animal behaviour. But evolution has travelled in the opposite direction. It is the tendency to preserve a stationary state which, together with the other more or less similar affective tendencies (to be considered in the next chapter), gives rise to the non-mechanized process of 'trial and error,' of 'provando e riprovando' which the *Accademia del Cimento* chose for their motto, not knowing that the humble infusoria had forestalled them in practising it. And it is the successive and progressive mnemonic fixation of the successful trial which transforms the latter into a reflex or a mechanized instinct.

On this point Bergson, apart from his untenable metaphysico-vitalistic theories, was perfectly justified in his criticism of Spencer: "By compounding the reflex with the reflex, Spencer thinks he generates instinct and rational volition one after the other. He fails to see that the specialized reflex, being a terminal point of evolution just as much as perfect will, cannot be supposed at the start." —"At the lowest degree of the animal scale, in living beings that are but an undifferentiated protoplasmic mass, the reaction to stimulus does not yet call into play one definite mechanism, as in the reflex; it has not yet choice among several definite mechanisms, as in the voluntary act; it is, then, neither voluntary nor reflex, though it heralds both."[1]

The same holds for the most complex instincts. They arose phylogenetically as non-mechanized behaviours, were supported by corresponding affective tendencies, proceeded to each new stage of their development by way of repeated trials, profited in each renewed attempt by the past experience of the individual and of the species, were gradually mechanized as the result of mnemonic accumulation, and thus finally became fixed or crystallized in the intricate morphological structure of the locomotor

[1] Bergson, *Creative Evolution*, London, 1911, pp. 394, 395.

apparatus and the nervous system. This structure, as such, like every other acquired morphological modification, can be transmitted indefinitely to succeeding generations and thus instincts may become inherited.

It may seem needless to insist on the purposive aspect of instincts, since it is by this that naturalists have been most impressed. But although the more complex instincts have attracted the most attention, yet even those which seem the simplest exhibit such a highly purposive co-ordination of movements that they appear no less marvellous. Hering, for instance, very rightly quotes the apparently simple instinct of the chicken which is able to peck grains of corn as soon as it comes out of the egg: "Without hesitation it picks up the grains which are thrown to it. This implies that it sees them, and that it correctly judges their position and their distance; moreover, it has to move its head and its limbs with great precision. All these things could not be learned in the egg-shell."—"The gentle stimulus of the rays which proceed from a grain and fall upon the retina of the chicken gives rise to the reproduction of a complicated series of sensations, perceptions and motions, which in this individual have never as yet been combined, and which, nevertheless, were adjusted from the beginning with accuracy and precision, as if the animal itself had practised them thousands of times."[1]

When we leave the simple instincts and pass on to the more complex, even the untrained observer cannot fail to see how closely their purposive nature resembles that of intelligent actions, in the strict sense; the only difference being that instincts seem to be already instructed by experience, while acts guided by intelligence have to learn

[1] Hering, *Ueber das Gedächtniss als eine allgemeine Funktion der organisierten Materie*, pp. 26–27.

from new experience: "The most obvious difference," writes Johnstone, "is that the instinctive action is effective the very first time it is performed, while the intelligent action only becomes effective after it has been attempted several times, or very many times, according to its difficulty."[1]

But instead of dwelling upon the purposive character of instincts, whether simple or complex, we shall find it more profitable to consider their strongly marked mnemonic nature.

The credit for having insisted on this point in a more suggestive manner than anyone else falls to Semon. He quoted, for example, Claypole's statement that young ostriches which have been bred in the incubator will only pick up food if the ground on which it is scattered is first tapped. And he suggests that of all the explanations which have been given for this fact by far the most probable is that which attributes it to the mnemonic evocation of an act learnt by the species in the past and transmitted hereditarily to the young ostrich; the act is ordinarily called forth by the demonstration of pecking by the mother, and in this case by the touching of the food with the fingernail or a small stick, this being, from the point of view of sound produced, an external stimulus very similar to that provided by the mother.

The following example is still more typical: a pan of water was put in front of a young magpie about five weeks old which had been reared by the observer from the time when it was hatched. It made one or two pecks at the surface of the water, and then, *outside the pan, although in the past it had never been immersed in water*, it proceeded to go through all the gestures which a bird makes after it has bathed—ducking its head, fluttering its wings and

[1] Johnstone, *op. cit.*, p. 284.

tail, squatting down, and spreading itself out on the ground. " The strangeness of the bird's conduct," Semon remarks, " becomes intelligible when regarded in the light of engram-inheritance. The stimulus of the contact with water, slight though it was relative to body area, acted ecphorically."[1]

Now the mnemonic nature of instincts, which is illustrated in these and innumerable other examples—that is, the mnemonic nature of hereditary transmission of a particular morphological structure, gradually acquired by a species through mnemonic accumulations deposited during individual life as a result of the repeated performance of a single act—implies the equally mnemonic nature of hereditary transmission of every other morphological modification which is also gradually acquired by the species through mnemonic accumulations deposited during individual life as a result of the repeated performance of a single functional adaptation. It is thus sufficient to explain both the purposiveness of instincts and that of the complicated and ' zweckmässig ' structure of organisms.

But, incredible as it may seem, even the purposiveness of instincts has been denied—for good reasons of their own—by extreme mechanists. To explain it away they once again call in chemical agents—hormones—which would at once ' produce ' directly in all its particular modalities each instinctive act.

" The idea," writes Loeb, " that the organism as a whole cannot be explained from a physico-chemical viewpoint rests most strongly on the existence of animal instincts. Many of the instinctive actions are ' purposeful ' (his use of inverted commas shows that he himself does not accept this interpretation), that is, assisting to preserve the individual and the race. This again suggests ' design'

[1] Semon, *op. cit.*, *The Mneme*, 1921, pp. 70–71.

and a designing 'force,' which we do not find in the realm of physics. . . . But when it can be shown that these instincts also may be reduced to elementary physico-chemical laws the assumption of design becomes superfluous."[1]

As usual he falls into the serious error of confusing the *physico-chemical agents which release instincts* (simple 'ecphoric' stimuli, as Semon would say) with the factors which directly produce and fashion them. "In the curious experiment of Steinach on the reversal of the sexual instincts of the male whose testes had been exchanged for ovaries," he writes, "there is little doubt that the sexual activities of each sex are *determined* by specific substances formed in the interstitial tissue of the ovary and testes."[2]

On the contrary it is perfectly clear that this is a case of mere release of instinctive tendencies. Both sets exist potentially in the organism, and either can be brought into activity according to the ecphoric stimuli presented. Loeb, following this method of reasoning, might even assert that the smell of seaweed *causes the image* of the Gulf of Naples, although obviously it only releases the memory of it.

If the countless acts which go to make up even the simplest instincts—the spinning of a cocoon by a silk-worm, the construction of a honeycomb by bees, the building of a nest by birds—were each caused directly by physico-chemical agents, then a much more formidable and inexplicable problem would arise—namely, how can the fortuitous play of so many physico-chemical agents, each working independently, lead to such co-ordinated results and produce such a definitely purposive aspect? By refusing to accept our hypothesis that there is in life a well-defined elementary property which differentiates

[1] Loeb, *op. cit.*, pp. 253, 254.

[2] *Id.*, *ibid.*, p. 254.

it from everything in the inorganic world and which can explain by mnemonic accumulation and transmission the most complicated instincts, mechanists are thus again obliged to assume the 'clock-maker' of Voltaire in order to explain how such a perfect mechanism came to be conceived and constructed.

The matter of this chapter can, then, be summarized in the following propositions:

1. The purposive aspect of the reflexes is so pronounced that they, more than any other manifestation of life, have given rise to the impression that the organism is a "machine."

2. The mechanists, however, forget that every machine is made *for a given end*, and that it presupposes a mechanician who has designed and constructed it.

3. We cannot, therefore, escape the dilemma: either we must grant Voltaire's 'clock-maker,' who is responsible for this machine, or have recourse to some fundamental property of life to take his place.

4. Far from being the basic phenomenon from which we must start in order to explain non-mechanized animal behaviour, the reflex represents only the gradual mnemonic fixation of a successful trial which a corresponding affective tendency has by chance produced and chosen. We cannot, therefore, understand the inner nature of animal behaviour, unless we understand that of the affective tendencies which we shall examine in the next chapter.

5. The purposive aspect of even the simplest instincts is so obvious that we cannot fail to notice it unless we allow prejudice to blind our eyes.

6. Instincts are of the same nature as intelligent acts, except that they are already informed by past experience while the latter are not.

7. For this reason, and because of the plentiful evidence

they furnish of being capable of being 'evoked,' the instincts appear to be of an essentially mnemonic nature.

8. The essentially mnemonic nature of the instincts, that is the mnemonic nature of the process of hereditary transmission of the particular morphological modifications in which they are crystallized, is proof of the equally mnemonic nature of the process of hereditary transmission of the whole structure of organisms, which is so complex, so perfectly co-ordinated and so 'zweckmässig.'

9. The attempt of extreme mechanists to deny the purposiveness of instincts and to explain them in purely physico-chemical terms shows once more the blinding effect, so mischievous for science, of preconceived ideas, "salde come torre che non crolla," when they are held with an Olympian imperturbability which neither self-criticism nor opposing facts can ever shake.

CHAPTER VII

Purposiveness of the Affective Tendencies

Affective tendencies which aim at preserving the normal stationary physiological state of the organism. Affective tendencies which aim at 're-activating' or restoring physiological conditions or environmental relations of the past. Affective tendencies arising from the two preceding groups by transference or composition. Since the origin and nature of all these affective tendencies are mnemonic, their purposiveness is also of a mnemonic origin and nature. Hence man's tendency towards the expansion and intensification of his individual life. Nothing remotely resembling affective tendencies is ever exhibited by brute matter, which, by antonomasia, is called inert. Mechanists have not attempted to explain affective tendencies, but have confined themselves to denying their existence in animals and *excluding* those of man from their jurisdiction, thus renouncing a unitary view and explanation of all vital phenomena.

From observation of the behaviour of the various organisms from the unicellular up to man, we see that many of their acts, and the more important ones in particular, can be interpreted as manifestations of the tendency of the organism, which we have frequently referred to in the preceding chapters, to maintain or restore its stationary physiological state.

In other words, if we reserve the term 'affective' for that particular class of organic tendencies which appear subjectively in man as 'needs' or 'appetites' or 'desires,' and objectively both in man and animals as *non-mechanized* movements, completed or incipient, then a whole group of affective tendencies, thus defined, can be reduced to the single fundamental tendency of each organism towards 'physiological invariability.'

We dealt with this point in the first chapter of our *Psychology of Reasoning*, where the mnemonic origin and nature of affective tendencies are considered, and in our

Michonis lecture, before the Collège de France, on the purposiveness of life, where we took the affective tendencies as a starting point. Thus hunger, thirst, the desire for the optimum environment, the striving to resume a normal position in relation to the force of gravity, the need for elimination of waste matter which is harmful to the whole organism, the sexual instinct itself and, finally, the instinct of self-preservation are but so many particular manifestations of this general tendency, common to all organisms from the highest to the lowest, to continue in a stationary physiological state.

The most varied and complicated animal movements of approach or withdrawal, of attack or flight, of seizing or rejecting, have all, directly or indirectly, the one exclusive end of maintaining or restoring the normal physiological invariability of the particular organism.

But, as we saw in the works we have just quoted, to this first group can be added a second, comprising all the needs or appetites or desires which spring *from habit*. In this category comes the craving which is felt for symbiotic or parasitic relationships after they have persisted for a certain length of time—as, for instance, those of mother and child from which maternal affection springs. Family affections in general, the ties of friendship, social sentiments, all the needs we acquire during life as the result of fixed habits or of some accustomed relation to environment and all the various nostalgias fall into the same group.

We proceeded to distinguish a third and last group—that of the *derivative and composite* affective tendencies which arise out of those of the two preceding groups, either by affective transference (from the whole to the part, from the end to the means, from a certain environmental relation to a part of it, from a certain object to another analogous one, and so on: that is, Ribot's *law of trans-*

ference), or by composition of two or several affective tendencies which come into action simultaneously and which, by association, interference, mutual reinforcement or partial inhibition of each other, give rise to a complex resultant affectivity, which, according to the number, quality and intensity of its component elements, is capable of exhibiting all the infinitely varied and delicate nuances which human sentiments can assume.

Finally, we succeeded in showing that the two first groups (of which the third, we repeat, is only a derivative) are themselves derived from the property of mnemonic accumulation which is the fundamental property of all living substance. In other words, all the affective tendencies (which are of an unquestionably teleological character) are due to the specific accumulations of corresponding physiological activities, provoked in the past by external circumstances.

The fundamental organic affective tendencies, that is all of the first group and some of the second, had their origin and seat, at first, in the soma or in some part of it; but as the nervous system developed, and particularly the brain in which all somatico-visceral activities are reflected, corresponding cerebral seats were added to and eventually substituted for the somatic seats. These purely cerebral seats are necessarily widely *extended* over the grey substance, in relation to the parts of the soma which they reflect. The remaining affective tendencies of the second group as well as the derivative and composite tendencies of the third, which are of a higher spiritual order, acquired similarly *extended* but exclusively cerebral seats.

It is because of their *widely extended* seats in the brain and because of the correspondingly greater potential energy which they represent that the affective tendencies, although in essence they also are pure mnemonic accumulations, differ so substantially from stored up sensory memories which are localized each in a single point of the brain and have a correspondingly small quantity of potential energy. In the *Psychology of Reasoning* we showed how important it is to take into consideration the great differences in extension of seat and in quantities of potential energy if the psychological differences between the play of released mnemonic accumulations, which are purely sensory evocations, and those which constitute affective tendencies are to be understood.

All these mnemonico-affective accumulations, therefore, represent so many accumulations of potential energy which, as such, tend to become active. But they are prevented from doing so simultaneously by the inhibiting influence which one exercises upon another.

From all of them and from their general tendency to become active arises that tendency *towards the expansion and intensification of individual life* which is typical of man, who alone among animals possesses such a wealth and variety of affective accumulations.

This tendency is essentially different from the tendency to the expansion of life in general, of which we spoke in a previous chapter, which is stimulated by solar or thermic energy. As a result of this external impulse the metabolic process draws into its vortex an ever larger quantity of brute matter and raises it to the level of organic substance. It acts more especially upon, and increases the production of, the germinal substance which, because of its greater autonomy and independence from the rest of the body, is particularly open to this external energetic influence. As a consequence there follows a heightened activity of the sexual instinct which tends to eliminate the germinal substance, thus augmented, in so far as it disturbs the normal stationary physiological condition of the organism. In the last analysis, therefore, it appears in the incessant multiplication, by generation, of the number of individuals, that is to say in the increase of population. This increase inevitably engenders Darwin's struggle for existence, if the population of all the species taken together or of any particular species exceeds the limit which is compatible with environmental conditions. The tendency towards expansion of life in general may thus be called *passive*, inasmuch as it is due to an external impulse; while the tendency towards expansion and intensification of individual life may be defined as *active*, since it depends not on external but on internal impulses, consisting of all the mnemonico-affective accumulations which, we repeat, only man possesses in such great wealth and variety.

Thus the most luxuriant forests are found in the Congo, Brazil, and Ceylon, that is in equatorial regions, whilst the most populous town in the world, the product of feverish human activity, is found in the north, in England. And whilst a single day's intense heat will cause mosquitoes to multiply by millions, the intense spiritual life of a genius, whether he be scientist or artist, is almost wholly independent of the thermic conditions of his environment.

We shall see in Chapter IX the importance from the point of view of social relationships of this active tendency towards expansion and intensification of individual life.

There is no need to insist here on the fact that we never see anything remotely resembling affective tendencies in brute matter, which, by antonomasia, is called *inert*.

Certainly, heavy bodies 'tend' towards the centre of the earth, two bodies charged with the same positive electricity 'tend' to repel each other, a fragment of cork or a similar floating body, when plunged into water, 'tends' to return to the surface, air 'tends' to rush into a vacuum, and two chemical elements with strong mutual affinities 'tend' to combine; but to explain these inorganic phenomena of impulsion or attraction or repulsion the physico-chemist has formulated the hypothesis (which is confirmed by the agreement between his deductions and observed facts) that shocks are produced between the respective bodies or that there are external forces in the intervening medium which cause these bodies to clash, to attract or repel each other and to which *they yield passively*. In the affective tendencies, on the other hand, whether of animals or of man, it is clear that the forces are *internal* and, in certain respects, independent of external forces or, at least, autonomous with regard to them.

Hence the *activity* and *spontaneity of movement* in the behaviour of living beings, in diametrical opposition to the *passiveness* and *inertia* of inanimate matter.

Hence, too, the fact that whilst we know exactly how inorganic processes will behave if we are acquainted with the external forces at work, unexpectedness is, on the contrary, characteristic of animal behaviour in so far as it is not mechanized in reflexes and instincts but is determined by the affective tendencies. This unexpectedness is due, in the first place, to the great number and variety of the affective tendencies, since the mode of behaviour depends upon the activation of one or other of the tendencies which cannot always be determined beforehand; and, in the second place, it is due to the fact that even if *the end* of any particular affective tendency which has been called into play is determined, the means which the animal will adopt to achieve that end when it finds itself in a new situation, still remain doubtful; for the choice may be fortuitous and dependent on the purely accidental result of a series of 'trials and errors' which the affective tendency in question has stimulated.

Nor need we devote much time to showing how extreme

mechanists have failed to suggest the slightest explanation of these indubitably purposive manifestations, but have confined themselves to a *denial* of the existence of affective tendencies, in animals, by reducing their whole behaviour to a complicated play of tropisms, while, as far as man is concerned, they have *dismissed* them and all other facts pertaining to the mind as falling outside their sphere. This was, in their opinion, the only hope of saving their point of view, because it was impossible even for them, by observing introspectively their own conduct, to deny that they are always impelled by some 'end' to be attained and that their needs and their desires are of an essentially teleological nature.

Without going into further details we can now summarize the conclusions of this chapter in the following fundamental propositions :

1. The purposiveness of a first group of affective tendencies is demonstrated by the fact that they are only so many particular manifestations of the general tendency, common to all organisms, to preserve their normal stationary physiological state unchanged.

2. The purposiveness of the second group of affective tendencies consists in their efforts to reactivate or restore certain physiological states or certain habitual environmental relations of the past, for which the organism experiences a longing (nostalgia).

3. The purposive nature of the first two groups is carried over to a third group which comprises affective tendencies derived from the former by transference or by composition.

4. Since the affective tendencies of the first two groups, and consequently also those of the third, are mnemonic in origin and nature, their purposiveness also is mnemonic in origin and nature.

5. The great diversity which exists both from the point

of view of the extent of their cerebral seats and from that of the quantity of potential energy, explains the differences, so fundamental from a psychological standpoint, which exist between the play of mnemonic activations constituting affective tendencies and the play of mnemonic activations representing sensory evocations.

6. The whole of the mnemonico-affective accumulations gives rise to the tendency towards expansion and intensification of individual life which is peculiar to man.

7. Nothing in the slightest degree resembling the affective tendencies is found in brute matter which, by antonomasia, is called *inert.*

8. The 'tendencies' of inorganic bodies are due to external forces to which they merely yield; the affective tendencies of animals, on the other hand, are due to internal forces, independent of external forces and autonomous with respect to them.

9. Hence the *activity* and *spontaneity* of the behaviour of living beings as opposed to the *passiveness* and *inertia* characteristic of brute matter.

10. The mechanists have not even attempted to explain the affective tendencies, but have confined themselves to *denying* their presence in animals and *excluding those of man from their jurisdiction. They thus abandon the unitary conception and explanation of all vital phenomena.*

CHAPTER VIII

PURPOSIVENESS OF MENTAL ACTIVITY

In the clearly purposive sphere of mental activity we have no opponents to reckon with, for mechanists have declared that it is outside their jurisdiction. The mnemonic theory, on the other hand, can account for it adequately and so enables us to attain to a unitary conception and explanation of all vital phenomena. This can be done simply by showing that all mental phenomena are derived from the reciprocal play of two elementary psychical phenomena, viz., the affective tendencies and sensory evocations, both of which are mnemonic in origin and nature. Will. Attention. The emotions. Consciousness and unity of consciousness. Imagination and reasoning. Coherence and logicalness. Power of abstraction and concepts. Abstract reasoning and mathematical reasoning. Dialectical reasoning and metaphysical reasoning. Affective activity, therefore, appears to us as the sole architect of thought and of all human psychical manifestations in general.

WE now enter a field which is clear of opponents, since the mechanists have completely abandoned it and declared that it falls outside the physico-chemical sphere. By so doing, we repeat, they have implicitly renounced a unitary conception of vital phenomena. The teleological character of this department of life is too evident to need demonstration, for every thought, every activity of the imagination or the reason, every meditated and voluntary action is always directed towards an end which mental activity itself seeks to attain by direct or indirect means.

But in order to prove that our mnemonic theory, in contradistinction to mechanistic physico-chemical theories, admits of a unitary explanation of life, we must rather show how all mental phenomena—even the most complex and lofty manifestations of the so-called spirit—can be deduced from that property of mnemonic accumulation which we have postulated as the peculiar and exclusive property of life.

To do this we need only to show that all mental phenomena, even the most complex, result from the reciprocal play of two elementary psychical phenomena, the affective tendencies and sensory evocations. Both these latter are of mnemonic origin and nature. If, therefore, all mental activity can be shown to derive from their reciprocal action, then it will be clear that this, too, in the last analysis, is derived from the same mnemonic property.

In our work on the *Psychology of Reasoning* we tried to show that this can be done.

Thus, for example, from the conflict between an affective tendency directed towards a future end and another directed towards a present end the *will* arises. From the conflict between a primary affective tendency which aspires to a certain good and a secondary affective tendency which holds the first momentarily in suspense for fear that if it were too free to come into play it might fail to achieve the desired result, *attention* emerges, with its attendant effects : more accurate perception and thorough examination of the particular object of interest at that moment. Again we saw how the too sudden and too intense activation of the affective tendencies gives rise to the *emotions*, while the abatement of this suddenness and intensity evokes the whole gamut of human *sentiments* up to *cool tenacity in action* which is affective but not emotive, and is the more efficacious according as the portion of the affective discharge dispersed in chaotic and useless visceral commotions accompanying the emotions is the less.

Consciousness, which so far has constituted the greatest enigma of psychical life, appears to us not as an intrinsic and absolute property inherent in each of the various psychical states, considered apart from the others, but a property extrinsic and relative to each one of them, due to certain modalities of affective reference existing between them. And *unity of consciousness* seems to be due to the great extension of the respective seats of the affective tendencies in the brain, which makes it impossible for more than one affective ' constellation ' to be in action at any one moment (excepting in the normal cases of *absent-mindedness* and the pathological cases of *double personality*).

Reasoning would appear to be nothing but a concatenated series of experiences which are merely thought, and the reasoner who ' thinks with attention ' seems moved at the same time by a primary and a secondary affective tendency: the primary, by means of suited sensorial evocations, imagines and follows ' with the mind's eye ' the various experimental combinations or vicissitudes to which the particular object of attention is subjected, whilst the secondary holds the primary in suspense at each stage in the process of attentive reasoning, for fear of attributing to each merely imagined experience a result other than that which would follow if the experience were actually accomplished.

We then saw how the *coherence* or *incoherence* of the whole intellectual process depends on the greater or smaller capacity for persistence of the primary affective tendency, whilst the *logical* or *illogical* character of reasoning itself depends on the greater or smaller capacity for persistence of the secondary affective tendency. This is confirmed first by the study of *dreams* which are both incoherent and illogical because of their non-affectivity; and secondly by studying *the reasonings of the insane*, which in mono-affective maniacs are coherent but illogical, while in the insane characterized by instability, impotence or absence of the affective tendencies they are remarkable for their incoherence.

From affective classification we saw that the *concepts*

are born, and these in their turn provide an increasingly solid basis for reasoning in its double evolution from concrete to abstract and from intuitive to deductive reasoning, which is proper to science, until the highest expression of thought is reached in mathematical reasoning.

Objective and constructive reasoning, or reasoning in its proper sense, is impelled solely by the desire to know and discover the truth, that is to foresee the consequences of certain actions before they have been embarked upon. From this we saw that little by little another type of reasoning, viz., *intentional reasoning*, becomes differentiated and then completely separated ; and this can itself be divided into dialectical and metaphysical reasoning. Of these the first aims at presenting certain given facts as belonging to certain given categories rather than to any others, while the second is concerned with making the whole universe appear as we wish it to be rather than as it actually is.

We thus saw that affective activity seems to permeate all processes of thought. And more than that, it seems to be the sole architect of thought, using the memories of the external world which are stored in our sensory mnemonic accumulations to build up all the constructions of the imaginative and ratiocinative faculty, from the lowest and meanest thoughts of the dull-minded to the most superb and sublime systematizations of the man of genius.

We thus see in the whole functioning of the intelligence not only the most typical illustration of the purposive aspect of life, but also the irrefutable proof of the complete inability of mechanism to explain life in its entirety and to win to a unitary conception of it.

CHAPTER IX

Purposiveness of Social Manifestations

JUSTICE AND MORALITY

Social purposiveness as the resultant in which the purposes of the individuals composing society are reflected and compounded. With the progress of social development the causes of conflict between individual aims themselves and between individual aims and collective aims increase on the one hand. But on the other hand the collective urge towards the assuagement and reconciliation of these conflicts becomes more imperious and irresistible. Thus arise the two great problems of social purposiveness which are peculiar to human society: justice and morality. Justice seeks to prevent, weaken and reconcile contrasts *by external adjustment*. Morality tends towards the same end, but *by purely internal adjustment*. In this way in place of the struggle for life there is an increasing tendency to substitute the joyous harmony of life, the ultimate goal and the supreme crowning point of biological evolution.

The fact that an essay on the purposiveness of life should also take into consideration the purposiveness of social manifestations ought not to cause surprise. Society, envisaged as the sum or resultant of manifold individuals, must reflect in one way or another the purposiveness of these individuals; and, inversely, the purposiveness of social manifestations will furnish us with fresh proof, *ad abundantiam*, of the purposiveness of life.

When the purposes of the individuals making up a whole converge, that is when the whole represents only the sum of its parts, the purposive social manifestations do not substantially differ from those of each part. Thus, for example, we see several men co-operating in the building of a house and taking several months over the work, while the savage does the same thing alone in the case of his wretched hut; but the ultimate purpose of both is the same—to make a shelter which will one day serve to protect them against the inclemencies of the weather. Again we see, in place of the individual housewife of old, large numbers of workmen working together in factories to produce, for example, the cloth which will enable the body, even in the depths of winter, to be kept in a constant high temperature, an indispensable condition for maintaining the invariance of the stationary physiological state. And, finally, we see town councils and states undertaking works of public utility, that is responding to needs which are felt by all citizens alike.

But we find a different state of affairs when the particular aims of some members of the community are in opposition to, or perhaps even in violent conflict with, those of other members of the community, notwithstanding that for both sides, since they cannot lead an isolated existence, it is a matter of life and death that they continue to form part of the same civil society in town or state. And the position is changed again when the sum of all these individual aims, more or less in harmony or conflict with each other, produces a resultant collective purpose which may be, and often is, opposed to certain particular individual aims which nevertheless enter into the composition of the collective purpose.

In the course of its evolution society has produced two orders of social facts which are diametrically opposed to each other.

On the one hand, with the growing complication of society, there has been a great increase in the motives of conflict between different individual aims and between individual and collective aims.

Thus, for example—to limit ourselves to the economic sphere—from the division of labour which, in our civilized societies, causes the production of a particular commodity (corn, wool, etc.), which is essential for the satisfaction of the needs of all members of the community, to be confined to a small minority, arises a conflict of interests between the producers of this commodity and the consumers who form the whole community.

In the same way, the technical complexity of the process of production which, for the making of goods at any one moment, requires not only the labour of that moment but the products of past labour, as raw materials and machinery, gives rise to a conflict of interests between the various classes of society: in particular between those who have only their labour to offer and those who are in a position to exploit the former at will, since they possess the products of the past which are necessary for the production of new goods.

And again, as a result of this division of labour, competition springs up between the producers of a particular commodity or the performers of a particular service. Each strives to secure for himself the greatest possible number of consumers for his goods or his services, for on the number of these depends the extent of his remuneration and on this his ability to satisfy his needs and desires. Those who emerge successfully from this feverish competition are very few, whilst the number of those who are crushed, who cannot even satisfy their basic needs, is great. The wretchedness and degradation of the latter in contrast to the excessive gratifications of the former are evidence of the gravity and the extension of the human misery which flow from this excessive and unorganized competition.

But, on the other hand, simultaneously with the multiplication of causes of conflict there has arisen a collective aspiration towards an assuagement and reconciliation of these conflicts, towards a substitution of harmony for the discords of life. This is due to the ever-growing intensity and extension of social relations and to the consequently more continuous and intimate communion of affective and spiritual life among members of the same collective group, so that in the finer spirits in whom this communion is fullest the aims of others have become their own.

Thus are generated the two great problems of social purposiveness which are exclusively proper to man: the problem of justice and that of morality.

Justice consists in seeking social constitutions and institutions which are able to contain individual activities and direct them into proper channels so that they may come into collision as little as possible; so that if conflict is inevitable, it may be mitigated as much as possible; so that individual aims may diverge as little as possible from the collective, or, better still, converge with it; so as, in short, to produce, *by external adjustment*, the greatest possible harmony between the various aims of individuals and between these and the collective aim.

This is a formidable problem which has troubled human societies from their first rudimentary manifestations in the taboo and the *lex talionis* to the most perfect juridical elaborations of our civilized communities. Through the philosophy of law in its widest sense it involves the whole study of the state and its fundamental institutions, including the different forms which the right of property may take, and all civil and penal legislation; it embraces the study of the various different social and political systems —autocratic, oligarchic and democratic—of the different forms of government, and the varying relations between the individual and the state. At the one extreme there is the excessive individualism which leaves the collective end or interest unprotected against the chaotic and anarchical aims of the various members; at the other extreme, a stifling idolatry of the state which abandons individual aims to the tyranny of the collective interest, which, moreover, is seldom the interest of the community as a whole but that of restricted privileged castes who dominate and exploit the others.

In the practical sphere, under the constant pressure of the different oppressed classes, castes or groups, this problem has impelled society towards an increasingly fuller correction of the most crying social injustices and inequalities and towards the progressive substitution in all social relationships of the system of free contract, the pre-eminent reconciler of individual interests, for the system of coercion which inevitably tramples on the interests of the many for the advantage of the few. In the realm of theory it has given rise to all the most idealized aspirations of humanity, from Plato's *Republic*, Campanella's *City of the Sun*, and More's *Utopia*, to the ideal of modern socialism which is the hope and faith of the present generation. And it was to the realization of this ideal that the author attempted to contribute, to the best of his then youthful abilities, in his work *Un socialisme en accord avec la doctrine économique libérale*, which he has since followed up and developed in subsequent secondary publications (v. *The Social Significance of Death Duties*, 1925).

The second problem, that of morality—to which in his future studies the author proposes to devote all his energies, now no longer invigorated by the freshness of youth, but still inspired and sustained by the hope of a happier future—seeks how best to mould the affective psyche of individuals so as to guarantee that the maximum number of individual purposes shall, *by purely internal adjustment*, be harmonized with each other and with those of society. In other words, the nature of this problem consists in striving to discover the supreme ethical postulate to be inculcated in the affective human psyche, the postulate from which may be deduced all those particular moral precepts which secure the spontaneous and harmonious adjustment of the vital activities of individuals in all their relationships with one another.

In this way instead of the activity or satisfaction of some individuals inhibiting and arresting that of others, the various activities and aspirations of all members of society might be given full and free play, even stimulating and intensifying one another with a maximum of activity and pleasure in life and a minimum of restraint and pain.

Thus harmony of life would take the place of the struggle for life.

As we saw above in the chapter dealing with the affective tendencies, it is only the brute beasts which submit passively to the external impulsion towards the expansion of life, which shows itself chiefly in an increase in the number of individuals and which inevitably fosters the struggle for existence whenever the population of all the species or of any single species exceeds the limit which is compatible with the conditions of the environment. Man, on the contrary, at any rate in his morally highest representatives, is capable of curbing and guiding this external impulse at will, so as to confine human population within the limits of available subsistence—limits, moreover, which are being continually pushed back by marvellous technical developments so that there is always a wide margin left. And we have seen how man substitutes for the *passive* acceptance of this external impulse a whole series of internal impulses based on the mnemonico-affective accumulations, which he alone possesses in such abundance and variety, and which together give rise to the really *active* tendency towards expansion and intensification of individual life.

This active tendency towards expansion and intensification of individual life may either continue to engender a wild beast struggle or else resolve itself into a harmony of life. Everything depends upon the quality of the affective accumulations and, above all, on their innumerable and infinitely varied modes of combination, composition, transference and transformation. And there is no doubt but that under the concerted impulsion of all the particular ends in which each individual seeks his gratification, this harmony of life will eventually become, not merely in theory but in practice, the sole and proper end of society as a whole. It is the business of affective education, on the one hand, and of the collective sanction, on the other, to stimulate and inculcate in an ever-increasing number of individuals those affective transferences, combinations, compositions and sublimations which weaken conflicting egoisms and transform them into harmonious ego-altruistic and altruistic affectivities. And it is the task of a moral genius, whether he be a Socrates or a Jesus of Nazareth, first to realize in himself and afterwards to stimulate in others, affective combinations which have never before been produced in the human psyche, and to discover and proclaim new moral postulates such as were never on men's lips before, and which mark so many forward stages in the arduous progress of humanity towards its radiant goal of joyous universal harmony of life.

We must not allow the doubts of the sceptic who points to the wickedness of the present time to hold us back or damp our enthusiasm in this vision of a better and nobler humanity. From the first metabolic attempts of the primeval unformed protoplasm towards its physiological stability to the whole purposiveness of life with all its infinite variety and splendour, culminating in the noblest ideals of the morally superior man, the path has been immeasurably longer than that which still remains for human society to tread in order to pass from its present imperfect attempts to master and soften the harshness of selfishness and hatred to the full triumph of its own particular purposiveness—a triumph which has already been attained by a chosen few, but which can only be completely realized when love and brotherhood reign among mankind.

And as the light in the radiant eyes of Beatrice drew up the body of the Poet through the heavens, so may this radiant vision of the joyous loving humanity which is to be, draw us with faith and enthusiasm along the rough path of goodness and duty which we have yet to tread in order to translate this vision into reality.

PART TWO

An Intermediate Solution Between Vitalism and Mechanism

CHAPTER X

VITALISM AND MECHANISM

Preliminary investigation completed. Purposiveness the rock on which the mechanist theory comes to grief. Mechanists recruited for the most part from the ranks of physiologists ; vitalists from ontogeneticists and psychologists. Both mechanists and vitalists recognize the necessity for a fundamental division of natural phenomena ; but they disagree as to where the dividing line shall be drawn. The method of investigation which the physiologist pursues prevents him from seeing the inner nature of vital phenomena. Vital activity, which connects physico-chemical phenomena, completely escapes the notice of the mechanistic physiologist. To employ only physico-chemical analysis is equivalent to excluding from the field of research the most fundamental aspects and problems of life. In its turn, however—which is even more serious—animistic vitalism entails the abandonment of any attempt to explain the nature of life. The vitalist theories of Driesch and Bergson are the fullest demonstration of the complete inability of animistic vitalism to explain anything.

I

Preliminary investigation completed

IN the Introduction to this work we said that before attempting to pronounce any judgment in the age-long dispute between vitalists and mechanists as to the nature of vital phenomena we ought first to consider the peculiar fundamental property common to all living organisms, viz., the purposive or teleological aspect of the most typical manifestations of vital processes. And, in addition, we said that the attitude we adopt towards this great controversy must finally rest upon this fundamental property.

We have attempted to perform this task in the preceding chapters in which the various purposive aspects of life have been passed in review, beginning with the most elementary biological phenomena and proceeding to the more complex and finally reaching those of mental and social life. Our examination of these various purposive manifestations, rapid though it was, has shown us that they can be satisfactorily explained by the hypothesis which postulates the existence of a special form of energy at the basis of life. This hypothesis provides us with the long sought for intermediate solution of the protracted controversy.

II

Purposiveness the rock on which the mechanist theory comes to grief

It was obvious that the long controversy between the vitalists and mechanists would end in an intermediate solution rather than in the overwhelming victory of one, for each party has excellent and irrefutable arguments with which to undermine its opponents' position. A way out of the dilemma there had to be, and it was only waiting to be looked for.

Sooner or later, indeed, it was inevitable that the mechanist theory should come to grief over the very difficult problem of purposiveness, which is now so irrefutably proved *by facts*. The mechanists themselves were perfectly well aware of this, and hence all their efforts have been directed towards *denying the existence* of these purposive manifestations of life. Thus Loeb has suggested the possibility that these manifestations are nothing more than a fallacious appearance : " The whole animated world is *seemingly* a symphony of adaptation."[1] But then the vitalists might well ask why no phenomenon in the inorganic world ever even *seemingly* possesses this characteristic.

Loeb has even gone so far as to deny the faculty for

[1] Loeb, *op. cit.*, *The Organism as a Whole*, p. 341.

adaptation: "An investigator examining the salinity of water and not knowing the natural resistance of Fundulus to change in concentration would be inclined to assume that he had before him an instance of a gradual adaptation of a fish to the higher concentration of the sea water; whereas the fish was already immune to this high concentration before coming in contact with it."[1]

But, as we saw in an earlier chapter, the experiments of Dallinger and Davenport have proved, on the contrary, that infusoria and other lower organisms can gradually become adapted to higher temperatures and higher concentrations of salt solutions, which would have *killed* them if they had been brought into contact with them suddenly. These facts and many others relating to new adaptation, with which we dealt in the chapter mentioned above, and especially the response to toxins by the production of antitoxins, establish irrefutably the reality of this power of adaptation which is inherent in life. It is, therefore, very significant that in order to defend his extreme physico-chemical theory Loeb finds himself obliged to deny the existence of this faculty, *in spite of the most firmly established facts.*

But Loeb is not the only one who is guilty of this error. Bayliss, for example, denies that the responses of living organisms to changes in their environment are always *the most appropriate,* that is *the most conducive* to the well-being of the individual. And again Rabaud, who is the most prominent representative of the mechanist school in France to-day, goes to much trouble to prove that the wonderful perfection of certain instincts is a pure legend.[2]

As if these quite undeniable imperfections of structure, reaction and instinctive behaviour were a sufficient

[1] Loeb, *ibid.*, pp. 328–329.

[2] W. M. Bayliss, "Vitalism," *Scientia,* April, 1922; and E. Rabaud, "Prescience et causes actuelles dans l'instinct," *Scientia,* June, 1920.

argument against vital purposiveness! We might just as well deny purpose to a machine which is not in perfect working order, or refuse to acknowledge the purposive tendency in human actions whenever they are not crowned with complete success.

III

Mechanists are recruited for the most part from the ranks of physiologists; vitalists from ontogeneticists and psychologists

It is a noteworthy fact that mechanists are recruited for the most part from the ranks of physiologists who work upon the already developed structure of the adult organism and who are not concerned with the way in which it has constructed itself nor with its psychical manifestations. But as MacBride says: "We have not only to explain how a finished machine works, but in addition how a machine builds itself up." And similarly Mathews writes: "The physiologist leaves to the psychologist all the psychical side of man's life. He takes, as a rule, no interest either in embryology or in evolution. It is only the men who consider these great problems of memory, inheritance, evolution and consciousness who become vitalists or who believe that it is not possible to explain vital phenomena on a mechanistic basis."[1]

The case of Claude Bernard is of great interest in this respect: when he is dealing as a physiologist with the fully developed organism he adopts a mechanistic point of view, but, by a flagrant contradiction, when he comes to deal with ontogenetic development he confesses himself a vitalist and recognizes that the embryo must necessarily be impelled by some 'directing force,' some 'invisible

[1] E. W. MacBride, "Vitalism," *Scientia*, July, 1922, p. 16; A. P. Mathews, "The Mechanistic Conception of Life," *Scientia*, October, 1924, p. 246.

guide ' along the path which it follows. " Even though we admit," he wrote, " that vital phenomena are connected with physico-chemical manifestations, which is true, yet very little light is thrown on the heart of the question by that admission : *for it is no mere chance encounter of physico-chemical phenomena which constructs every organism according to a fixed and pre-established plan and design, and which is responsible for the wonderful harmony and subordination of vital activities.*"—" Every organism and every organ seems as though it were constructed according to a pre-established design," he reiterates, " so that although, when taken in isolation, each part of the system appears to be in dependence of the general forces of nature, when taken in its relations with the others it shows a special connection, *it seems to be directed by some invisible guide along the road which it follows and led into the place which it fills.* The slightest reflection will make us realize that this pre-established regulation of life is a fundamental characteristic, a *quid proprium* of the living being."[1]

Mechanists are even more numerous among plant physiologists, simply because the much less intense nervous activity of plants weakens and obscures their mnemonic and consequently their purposive manifestations. But, as we saw in our earlier works, these manifestations are not absent in the vegetable world which exhibits sensibility, movement and mnemonic phenomena together with certain determined preferences, not to speak of metabolic activity which is in itself purposive.

Thus we see that where purposive manifestations are hidden or remain unnoticed mechanism still tries to hold out ; but where they are pronounced it is constantly yielding ground before the attacks of vitalists. " Dastre

[1] Claude Bernard, *Leçons sur les phénomènes de la vie commune aux animaux et aux végétaux*, Paris, 1878, pp. 50, 51.

asserted," says Piéron, himself an anti-vitalist, "that the last stronghold of the theory of a vital force was the sphere of morphogenesis, that is of development. It is true that here in particular contemporary neo-vitalism is trying out the strength of its explanations and the weight of its deductions. But there remains still another property of life for which it is difficult to establish mechanist hypotheses: not the building up, at a given moment, in favourable circumstances, of an organism, nor the realization of a particular function, but the organism's capacity for endurance and for resisting injurious influences and destructive forces. We see an ability to profit by experience, and changes in behaviour, in response to a harmful stimulus, which henceforth enable the organism to escape from this stimulus." And similarly Thomson writes: "The living being is in a category to itself because it possesses the power of recording experience, of storing up the past, of capitalizing its acquisitions; it is capable of self-conservation, of self-preservation; it acts always with a purpose and, in its higher stages, it is able to settle its own purposes."[1]

But, obviously, the phenomena of the human mind, which are so completely imbued with purposiveness, are those which present the greatest difficulty and which finally cause the breakdown of mechanist theories. "Mechanism," writes Conklin, "must account for purpose in man, as well as for fitness in lower organisms, if it is a universal principle." Dendy in his turn writes: "Can the mechanist explain thought as a physico-chemical process? Will chemistry and physics help him to explain why he loves his children?—It seems to me that if the mechanist cannot explain thought and love and psychical

[1] H. Piéron, *art. cit.*, "Du rôle et de la signification du conflit scientifique entre mécanisme et vitalisme," p. 123; J. A. Thomson, *art. cit.*, "Vitalisme méthodologique," p. 35.

manifestations in general by means of mechanistic principles he is faced with a dilemma: either he is compelled to admit that his mechanistic theory does not hold universally or he must try to save it by maintaining that man, at the very least, is outside his sphere."[1]

That is exactly what almost all physiologists do, for while they remain mechanists in their own particular sphere of investigation, they declare themselves spiritualists in the realm of thought; as though it were possible to draw a clear line of demarcation between physiological and psychical phenomena and as though the organic and fundamental affective tendencies which connect the two were not proof of the impossibility of separating phenomena of an essentially similar nature into two distinct classes. "As for the higher nervous functions, the mental functions," Piéron writes, "they present no difficulty to certain mechanists who abandon the whole of the sphere of thought to a spiritual principle and consider that the laws of matter cannot be expected to extend so far."[2]

IV

Both mechanists and vitalists recognize the necessity for a fundamental division of natural phenomena; but they disagree as to where the dividing line shall be drawn

The whole difference between mechanists and vitalists, therefore, is reducible to the following: the fixing at one point rather than at another of the precise dividing line which man finds himself compelled to draw between natural phenomena. The vitalists draw this line on the threshold of biological phenomena; the mechanists on the threshold of psychical phenomena. "In the course

[1] Conklin, *op. cit.*, *Problems of Organic Adaptation*, pp. 363–364 A. Dendy, *art. cit.*, "Mécanisme et Vitalisme," p. 327.
[2] Piéron, *art. cit.*, p. 122.

of the evolution of matter," writes Dendy (a vitalist), "a point is reached beyond which the relations of energy acquire a new aspect and a new significance which justify the application of the epithet 'vital' to them." Bayliss, on the other hand, a mechanist, places the great point of division "before the appearance on the scene of the conscious mental activities of man."[1]

If, in attempting to bring biological facts into the category of phenomena governed solely by physico-chemical laws, the mechanists had succeeded in unifying all the phenomena of the natural world, the persistence with which they pursue their end against all the most evident facts might be comprehensible and more justifiable. But, on their own confession, this unity always breaks at one point or another, and they themselves are the first to declare that the phenomena 'of mind' do not and cannot enter into physico-chemical jurisdiction. This being the case, it is far better to look with greater objectivity for the point where the line can in fact be drawn, instead of persistently denying with closed eyes that it must be drawn at such a point, only later to have to draw it a little further beyond.

V

The method of investigation which the physiologist adopts prevents him from seeing the inner nature of vital phenomena

Let us examine a little more closely, however, the reasons upon which, in the last analysis, mechanists base their denial that biological phenomena in general and physiological phenomena in particular are governed by laws other than purely physico-chemical ones. It is

[1] A. Dendy, *art. cit.*, "Mécanisme et Vitalisme," pp. 330–331; W. M. Bayliss, *art. cit.*, "Vitalism," p. 298.

evident that the very method of investigation which the physiologist employs, *which only allows him to take note of physico-chemical phenomena,* leads him to assert that these are the only ones which exist in life. "Physiological analysis," Johnstone writes, "shows, indeed, a resolution of the activities of the organism into chemical and physical reactions. How could it do otherwise? How could chemical and physical methods of investigation yield anything else than chemical and physical results? The fact that these methods *can* be applied to the study of the organism with consistent results shows that their application is valid; that we are justified in seeing physico-chemical activities in life. But are these results *all* that we have reason to expect?"—"It is difficult," he continues, "for the modern student of biology, saturated with notions of biochemical activities, gels and sols and colloids and reversible enzymes and kinases and the like, to realize that the belief in a vital agency is an intuitive one, and that the mechanistic conception of life is only the result of the extension to biology of *methods* of investigation, and not a legitimate conclusion from their *results.*"—And again he insists: "The physiologist adopts the methods of physical chemistry in his investigation and whatever results he obtains are necessarily of the same order. Inevitably, from the mere nature of his method, he can see, in the organism, only physico-chemical phenomena."[1]

If the thermometer were the sole instrument which physiologists possessed it is obvious that they would see in vital processes only rises or falls in temperature: in whatever part of the brain, for example, the ampulla of a thermometer is placed, only changes in temperature will be observed. But would the observer be justified in main-

[1] Johnstone, *op. cit., The Philosophy of Biology,* pp. 109–110, 120–121, 143–144.

taining that these variations in temperature are *all* that happens in the brain ?

Similarly a physiologist equipped only with instruments for the measuring of electric currents would be able to measure only the variations in these currents ; but that does not mean to say that when in the nerves, for example, he notes the presence of these electric currents alone, there is nothing else there. " When we think," writes Hering, " of the infinite variety of different processes which are able to produce electric currents, ought we not to hesitate before concluding from the identity of the electro-motor behaviour of two nervous fibres, the identity also of their nervous processes ? "[1]

Electrical phenomena are observable in all cellular activities ; but that does not mean to say that the whole of vital activity can be reduced to these phenomena. " Muscles, glandular cells, vegetable cells, possibly all living substance," continues Hering, " in all circumstances exhibit electrical phenomena which, in their mode of appearance, are strikingly analogous to those produced in the nerves : ought we to conclude from this that the internal chemical processes, which are the cause of these phenomena, are the same in the living substance of all these parts ? "[2]

If from the identity of thermal or electrical effects we cannot infer identity of the causes which produced them, still less can we learn from these effects the nature of the causes, or the essential content of the vital process of which they are, as it were, merely the collateral manifestations or posthumous products. To quote Hering once more : " In the electrical phenomena present in the nerves and muscles I have never seen anything other than *physical*

[1] Hering, *op. cit.*, *Zur Theorie der Vorgänge in der lebendigen Substanz*, p. 80.

[2] Hering, *ibid.*, p. 81.

symptoms which are as little able to enlighten us concerning the qualitative aspect of the processes taking place in the living substance as thermic phenomena are."[1]

It is for this reason that the method of investigation which is deliberately confined to the recording of purely physico-chemical phenomena (which, instead of emphasizing, rather prevent the special characteristics of life, and particularly its various purposive manifestations, from being seen) *has not taken us a step further towards a knowledge of what life really is.*

VI

Vital activity, the connecting link between physico-chemical phenomena, completely escapes the observation of the mechanistic physiologist

Vital activity is connected with physico-chemical actions and reactions inasmuch as it springs from and ends in them ; but this does not mean that it is identical in nature with them. It is *the intermediate connecting link* binding together the physico-chemical phenomena which are its starting point with those in which it ends. The physico-chemical physiologist sees the starting point and the finishing point, *but he does not see the intermediate link which connects them.*

The physiologist who persists in seeing in every manifestation of living matter only the physico-chemical points of departure and arrival *sees everything except life.* He fails to see the co-ordinating, controlling and directing power which the intermediate link, which alone really constitutes life, exercises over these physico-chemical phenomena. " Physico-chemical processes in the organism," writes

[1] Hering, *op. cit.*, *Zur Theorie der Nerventhätigkeit*, p. 106 ; and *Zur Theorie der Vorgänge* etc., p. 88.

Johnstone, "are only the *means* whereby the latter develops, and grows, and functions, and acts. In the analysis of these processes we see nothing but the reactions studied in physical chemistry; but whenever we consider the organism as a whole we seem to see a co-ordination, or a control, or a direction of these physico-chemical processes."—"Life is not only energy but also the direction and co-ordination of energies."[1]

Thomson's statement that "no vital operation *in its entirety* has ever as yet been described in terms of chemistry and physics" is perfectly justified.[2]

The 'inalienable unity' of the whole organism, due to vital activity, that is to the intermediate link which connects physico-chemical processes, disappears in physico-chemical analysis, and with it one of the most fundamental characteristics of organisms. "Even the simplest vital activity," Russell writes, "cannot be adequately and fully interpreted in all its relations as being purely physical; it has peculiarities which demand a psychobiological or functional interpretation. I do not see how adaptability, persistence of effort, co-ordination and unity of activities are to be understood save on the analogy of our own behaviour as living beings. The responses of living things, whether they be by movement, as in behaviour, or by metabolic change, as in growth and functional adaptation, must be considered as having an inalienable unity of their own, which is lost if they are analysed in terms of physico-chemical reaction."[3]

Now, as we have seen, everything justifies us in assuming that this connecting link, which actually constitutes life, is composed of the same nervous energy as that to which

[1] Johnstone, *op. cit.*, pp. 160, 340–341.
[2] J. A. Thomson, *art. cit.*, "Vitalisme méthodologique," p. 29.
[3] E. S. Russell, "The Question of Vitalism. Psychobiology," *Scientia*, September, 1924, p. 176.

the whole activity of the psyche is due. For this reason we are able by our hypothesis to furnish a single explanation which is equally applicable to biological and psychical phenomena and which brings out the essentially similar nature of the two ; we have thus succeeded in giving that 'psychobiological interpretation' of the facts of life which Russell demands.

VII

To employ only physico-chemical analysis is equivalent to excluding from the field of investigation the most fundamental aspects and problems of life

To make use only of physico-chemical analysis in biology implies the tacit exclusion of all the most fundamental aspects and problems of life which elude such analysis. Hering quite rightly says that the physico-chemical physiologist has hardly yet passed the threshold of the vital process : "To whatever point the physical or chemical investigation of the animal organism has penetrated, it has always, sooner or later, come upon the mysterious action of the living substance of those elementary organisms of which the animal and the human body are composed. We have now learned modesty, and from having once believed that we had entered the holy of holies, we now acknowledge that we have as yet scarcely passed the portico of the temple."[1]

Cease, therefore, advises this eminent physiologist, to see in physiology only applied physical chemistry. Physical chemistry, which is born from the study of the inorganic, can and must suffice to explain all that is common to living and non-living matter, but it is not sufficient to explain the special characteristics of life, that is those characteristics

[1] Hering, *Zur Theorie der Nerventhätigkeit*, p. 107.

by which a living being is differentiated from non-living matter. " Life," he writes, " can be *fully* understood only from life, and a physics and a chemistry which have sprung solely from the domain of inanimate nature, and which therefore apply solely to inanimate nature, are adequate only for the explanation of such things as are common to the living and the non-living."[1]

Russell develops the same idea: " It seems to me perfectly clear that the method of the physical sciences is impotent to take full account of those peculiarities of living things which obviously distinguish them from ordinary mechanisms—their enduring individuality, persistent through change, their power of profiting by experience and the like."[2]

" It is for this reason," concludes Hering, in agreement with Thomson, Russell and the majority of non-mechanistic biologists, " that psychology is an indispensable auxiliary science to physiology."[3]

If we look back, therefore, to what has been said in earlier chapters, we can readily answer the question which Huxley put, in 1869, in his famous lecture on " The Physical Basis of Life." He asked what were the reasons which justify the assumption that there exists in living substance something which has no representative or correlative in non-living matter—a question with which he thought triumphantly to confound non-mechanists. Our reply would be: the whole behaviour of living substance from the most elementary phenomena of assimilation and metabolism to the most complex psychical phenomena. Its constant purposiveness, which is completely different from anything that we see in the essentially non-teleological inorganic world, not only authorizes but compels us,

[1] Hering, *ibid.*, 107–108.
[2] E. S. Russell, *art. cit.*, pp. 173–174.
[3] Hering, *Ueber das Gedächtniss* etc., p. 8.

if we are really to understand the peculiar and fundamental properties of life, to attribute to it, even if only by way of hypothesis, a special form of energy, constituting the connecting and hidden basis of the physico-chemical phenomena which the physiologist can record with his physico-chemical instruments, but not explain. And this hypothesis will be completely legitimate if all the purposive manifestations of life receive from it the adequate explanation which till now has been looked for in vain.

VIII

In its turn, however—which is even more serious—animistic vitalism entails the abandonment of any attempt to explain the nature of life

Though mechanism has not succeeded and cannot succeed in explaining the essential properties of vital phenomena it at least represents *a positive attempt at explanation.* Vitalism, on the contrary, in the forms which it has assumed up to the present, represents, as th mechanists have quite rightly insisted, *not even the beginning of any such attempt,* but rather *the definite abandonmen of all explanation of life.*

In so far as vitalism confines itself to postulating a vague, undefined and nebulous entity as the cause of the purposive manifestations of life, and in particular of ontogenetic and psychical manifestations, it thereby implicitly affirms that these purposive phenomena *are inexplicable,* instead of offering an explanation of them.

" Innumerable attempts," writes Conklin, " have been made by philosophers and biologists to find an explanation for adaptation (finalism). One need only enumerate the 'supernatural design' of theologians, the 'perfecting principle' of Aristotle and Nägeli, the 'indwelling soul' of

Plato and Bruno, the 'active teleological principle' of Kant, the 'unconscious purpose' of Hartmann, the 'vital activity' or 'vitalism' of Bunge, Wolff and Virchow, the 'will' of Schopenhauer, the 'élan vital' of Bergson, the 'entelechy' of Driesch, the 'archæsthetism' of Cope, and so on, to indicate over what a wide field these attempted explanations have ranged."[1]

But it is clear that all these attempts, which are at bottom identical, come down to the introduction of a phrase, a *flatus vocis*, which merely succeeds in demonstrating but not in explaining the mysterious purposive properties of life; of them Mephistopheles might well have said: "Denn eben wo Begriffe fehlen, da stellt das Wort zur rechten Zeit sich ein" (just where concepts are wanting the word intrudes itself in good time).

All these animistic vitalist explanations and conceptions, based on a 'vital impulse' or a 'nisus formativus' or 'entelechies' and the like, destitute though they are of scientific value, are yet very significant, for, as Stefani has pointed out, they "implicitly recognize that not only have physics and chemistry not yet explained life, but that life cannot be explained by means of physics and chemistry."[2]

IX

The vitalist theories of Driesch and Bergson are the fullest demonstration of the complete inability of animistic vitalism to explain anything

When Driesch defines his 'entelechy' as a non-spatial, non-material, non-energetic agent, endowed with the power of regulating and co-ordinating the energetic

[1] Conklin, *op. cit.*, *Problems of Organic Adaptation*, pp. 329–330; v. also, for example, E. Gley, "Le Vitalisme. Evolution de la question," *Scientia*, July–August, 1923.

[2] A. Stefani, *Sul concetto della vita*, p. 8.

physico-chemical processes of the organism in such a way that they always lead to the "realization of an end," he really explains nothing. From the positive philosophical point of view to 'explain' is nothing other than to deduce by means of a series of imagined experiments unknown complex phenomena from simpler phenomena with which we are more familiar. Now this non-spatial, non-material, non-energetic metaphysical entity, endowed with intelligence and will, which conceals under a new name the old idea of the soul, is extraordinarily complex, as much and more so than the human psyche itself. Far from being the elementary phenomenon from which more complex phenomena such as ontogenesis and the whole biological and psychical behaviour of organisms can be deduced, it calls itself for explanation, even more than these other phenomena which it should explain.[1]

The more so in that intelligence, will, and the complex working of the mind in general, which are now beginning to be explained as deriving from the complex play of more elementary psychical phenomena, have their own spatial seat, their own material basis, their own energetic potentiality, constituted by the spatial seats, the material bases and the energetic potentialities of those elementary psychical phenomena of which the intelligence, the will and the complex working of the mind in general are composed and which, as we have seen, can be reduced to the affective tendencies on the one hand and to sensations and sensory evocations on the other. But in opposition to this, Driesch's 'entelechy' is represented, we repeat, as being an agent without a spatial seat, without a material basis and without energetic potentiality. Hence his comparison with the human intelligence and will, far from

[1] V. Driesch, *op. cit.* (*Der Vitalismus*), Part II, last chapter, "Die Lehre von der Entelechie im Wissenschaftsganzen," particularly pp. 235, 240, 242.

simplifying his idea, only tends to make it more obscure, since the 'entelechy' lacks all the spatial, material and energetic conditions which, according to the findings of psychology, are necessary for the functioning of the human intelligence and will themselves.

Driesch's entelechy, says Russell, "which is active in morphogenesis and restitution, is credited with faculties of *primary* knowing and willing, that is, of knowing and willing without previous experience." And more: "To all intents and purposes an entelechy corresponds to a psychical entity or soul, that is to say something which has perceptions without sense-organs, thoughts without a brain, and which acts without executive organs."[1]

No less tinged with mysticism is Bergson's theory. He sees life as "consciousness launched into matter," turned "either in the direction of intuition or in that of intellect," according as it "fixed its attention either on its own movement or on the matter it was passing through." This "consciousness launched into matter," this "vital principle," is called in to secure constancy of effect (*e.g.*, the specific form of the organism as the goal of ontogenesis) in spite of the wavering of causes (in the midst of which the organism develops).[2]

The purely verbal explanations of Driesch and Bergson, and others of the same kind, all more or less based on mystical and anti-scientific conceptions of a soul, separate and distinct from the material of the body (the tacit implication being its survival of the body), have discredited vitalism as a whole. And it is for this reason that physiologists to-day are suspicious, wrongly it is true, even of vitalistico-energetic hypotheses which are put forward

[1] E. S. Russell, *art. cit.*, "The Question of Vitalism. Psychobiology," pp. 170, 172. *Cf.* also, *e.g.*, Max Hartmann, *Biologie und Philosophie*, Berlin, 1925, pp. 37–38.

[2] H. Bergson, *Creative Evolution*, p. 191–192, 238.

by strict positivists and which have nothing in common with animistic theories.

It is the fear of being forced into vitalistico-animistic explanations, that is of being compelled to introduce an extra-phenomenal regulating intelligence, the admission of which is incompatible with all positive scientific thought, which leads mechanistic biologists to deny the purposiveness of life.

Nevertheless, as we have seen, the purposive manifestations of life are *facts*, and *indisputable facts*. Instead, therefore, of flying in the face of all evidence and persistently denying them, it is far better to seek to explain them, and to explain them *scientifically*, that is, by approaching them from the deterministic and causal point of view.

This requirement is fulfilled by our vitalistico-energetic hypothesis, which we shall examine in the following chapter.

CHAPTER XI

The Vitalistico-Energetic Solution

A causal and deterministic explanation of biological purposiveness furnished by our theory of life. The 'psychical stamp' of all vital phenomena is simply a mnemonic stamp. An objection: who has seen the nervous energy which is assumed at the basis of life? This objection shown on examination to be entirely worthless. It might equally well be denied that the ether of Fresnel has no value as explaining the phenomena of light transmission. It is animistic vitalism and not energetic vitalism which discourages scientific investigation. A new series of experiments suggested by our hypothesis. The accusations legitimately brought against all animistic vitalist theories, of running counter to determinism, to the principle of causality, to the foundations of energetics, in no way touch our vitalistico-energetic hypothesis. Our hypothesis concerning the nature of life stands, therefore, as a half-way theory which is capable of settling the age-long controversy between vitalists and mechanists.

Having considered the various purposive manifestations of life and examined the objections to the two opposing conceptions of vitalism and mechanism, we have now to draw the conclusion from these preliminaries and consider whether our intermediate hypothesis is free from the chief objections with which we have been dealing, and whether it can really claim to be the intermediate solution which will put an end to the long debate which still goes on between the two opposing conceptions.

I

A causal and deterministic explanation of biological purposiveness is furnished by our theory of life

Our mnemonic theory enables us, as we have shown at length in earlier works and more briefly in the first part of this book, to explain all the purposive manifestations of life, from the most elementary to the most complex, in a way which, though neither ' mechanistic ' nor narrowly ' physico-chemical,' is yet strictly ' energetic ' ; that is to say, *not animistic and metaphysical*, but, like the explanations of all other natural phenomena, *causal and deterministic*. Hence the recognition and acceptance of these purposive manifestations of life no longer involves the repudiation or disavowal of a positive *Weltanschauung*.

This result has been achieved, as we have seen, by means of the simple hypothesis of specific accumulation which rests on a well-defined energetic basis. According to this hypothesis, the mnemonic accumulations of those physiological activities which were determined in the past by environmental conditions, act, as specific potential energies, as an actual *vis a tergo*. It is this *vis a tergo* which, by impelling to the reactivation of those physiological activities of the past, causes the tendency towards the restoration, in one way or another, of the corresponding conditions of environment which are necessary for this reactivation, and which thus appear to act as a *vis a fronte*.

' Final causes ' are scientifically inadmissible when they are conceived as a ' future,' which obviously cannot have any effect before it exists. In biological finalism, however, it is not the *future* which is active but the past, through the mnemonic accumulations which it has deposited.

To explain all the purposive manifestations, which are so varied in their external forms yet so essentially similar in their inner nature, we have not found it necessary, in contradistinction to vitalist theologians and metaphysicians, to make use of the hypothesis of an " ordering intelligence which, as it were, invests organisms with activities which enable them to tend of themselves, blindly, towards the end which it assigns to them." Nor have we found it necessary to invoke "a co-substantial internal principle guiding and co-ordinating the single actions of the various parts of an organism with a view to the realization of an end," nor yet "an immanent principle in the organism, working towards an end, and realizing in the process of construction of the individual form a predetermined structural plan and nothing but this plan." These are nothing more than purely verbal expressions which make the phenomenon to be explained still more obscure.[1]

Again we have not, with Neumeister, been compelled to introduce any sort of 'psychical thinking' into the chain of physiological phenomena and to attribute to it "a directing action over the physico-chemical processes which take place in organized beings."[2]

To postulate at the basis of life such a 'psychical thinking' of which neither the nature nor the origin nor the method of functioning is known, and which is an unknown much greater than the facts which it is supposed to explain, is to fall into pure metaphysics and to renounce all positive scientific explanation. On the other hand, nothing can be more strictly scientific, nothing less in conflict with the most positive *Weltanschauung*, than to assume at the basis of life a special form of energy, endowed with a special, essentially elementary and well-defined property such as that of specific accumulation, from which can be deduced all the purposive manifestations of life, including those of the thinking psyche which are characteristic of mind.

[1] *Cf., e.g.,* A. Gemelli, *L'enigma della vita,* Florence, 1910, pp. 161 and 510.

[2] E. Neumeister, *Betrachtungen über das Wesen der Lebenserscheinungen,* Jena, 1903; quoted by Stefani in the paper mentioned above, p. 29.

II

The 'psychical stamp' of all vital phenomena is simply a mnemonic stamp

Because of the purposive tendency which they acquire from the property of specific accumulation, vital phenomena appear to have a marked psychical stamp. For this reason many biologists assert that psychical activity is present in every living part of every organism, whether animal or vegetable. Grassi, for instance, writes: "If we admit the existence of an unbroken chain linking man to other beings, if we admit that it would be irrational to dismember the kingdom of living beings, then we must also admit that some glimmering, however faint, of our unquestionably existing psyche is present in all the living parts of every organism, whether animal or vegetable."[1]

Now this "faint glimmering of the human psyche," whose presence we must recognize in the smallest fragment of living matter, is nothing other than the special form of energy which we assume as the basis of life and which alone gives to all its manifestations their purposive imprint. Similarly the 'vital impulse' of Bergson, the 'entelechy' of Driesch and all the other 'active teleological principles,' to which is assigned the part of *guiding* and *directing* the succession of physico-chemical phenomena in the organism towards given ends, instead of remaining vague metaphysical concepts, acquire a positive precision and consistency, if they are considered as nebulous substitutes for the specific accumulations postulated by our hypothesis, on the activation of which depends the progression of the organism towards corresponding 'ends.'

It is then the mnemonic accumulations which 'direct' the very phenomena produced by their activation. No

[1] Grassi, paper quoted: *La vita: ciò che sembra a un biologo*, p. 16.

separation exists, therefore, between the agents directing the phenomena and the phenomena directed, both being one and the same thing, viz., *mnemonic accumulations in the process of activation.* "A 'purpose' cannot work," as Driesch himself very rightly says; "the expression in question strictly means that something works 'which embodies the purpose in itself.'"[1]

As Mathews says, "it is precisely the phenomenon of memory which confronts the biologist at every turn"; but it is just this phenomenon that the physico-chemist never comes across in his work in the laboratory. Now the person who ignores memory and tries to explain merely the immediate happenings of an organism on a physico-chemical basis "is wilfully shutting his eyes before, or unconsciously neglecting, the most fundamental, the most characteristic, the most puzzling of vital phenomena."[2]

In other words, the true biologist cannot fail to consider the organism as a 'historic being,' as many writers have described it. Historic, not in the sense that its actual condition is simply a consequence of its past (for that could be said of all inorganic bodies), but in the sense that the past *continues to be equally active in the present,* in the form of mnemonic accumulations deposited by past activities, so that the *present* organism reflects and reproduces in itself the history of the past.

III

An objection: who has seen this nervous energy which is assumed at the basis of life?

But our critics will say: who has ever seen this nervous energy, this special form of energy which you postulate

[1] Driesch, "Le vitalisme," *Scientia,* July, 1924, p. 19.
[2] A. P. Mathews, *art. cit.*, "The Mechanistic Conception of Life," p. 244.

at the basis of life? It is perfectly true that no one has ever seen it, no one has ever touched it in his laboratory. How could it be otherwise since the physico-chemist has in his laboratory instruments which record only physico-chemical phenomena? If, instead of the galvanometer, the physico-chemist had only balances or thermometers or barometers would he ever be able to register an electric current?

Only when the physiologist has invented instruments capable of recording it, will nervous energy become perceptible to his senses. But meanwhile this energy can still be valuable as a hypothesis enabling the facts revealed by observation to be deduced and explained, exactly as the Newtonian force of gravity, which no one has yet seen or touched in the laboratory, serves as a hypothesis enabling certain terrestrial and celestial movements, recorded by observation, to be deduced and so far explained.

Fresnel made use of the hypothesis of ether, which no one has seen and no instrument can register, to explain the various phenomena of light; and this hypothesis is a true explanation in the strictest sense, since from the elementary well-defined properties which Fresnel attributed to ether we have been able to deduce all the most fundamental and characteristic phenomena of light. We are, therefore, justified in claiming the same right of recognition as an explanatory hypothesis for our theory which endows the nervous energy assumed at the basis of life with elementary properties which are equally clearly defined and which enable the most fundamental and characteristic phenomena of life itself to be deduced. The objection that the existence of such energy has not yet been verified in the laboratory is of no more force than would be the objection that no one has yet verified the existence of ether.

If an engineer should come down from Mars and confine

himself to the evidence furnished by sight, he would certainly deny the existence of the electrical current in a hydro-electrical mill and see, from the waterfall to the power supplied by the dynamo-motors, nothing but purely mechanical phenomena ; he would entirely miss the intermediate link, of electrical energy, which inserts itself between the initial phenomenon of the waterfall and the final mechanical phenomenon of the power furnished by the dynamo-motors. And the way in which the power has been obtained from the first mechanical phenomenon would remain an enigma, even though he might see that the phenomenon as a whole obeyed the fundamental law of the conservation of energy.

Similarly a Martian chemist who knew nothing of electrical phenomena would see only chemical phenomena in electrolysis ; he would equally fail to see the intermediate link of electrical energy which is the actual cause of the phenomena observed. For him, too, the way in which these phenomena are produced would remain an enigma, even though he should see that the laws of chemical composition and decomposition were not violated.

So it is with the physico-chemical physiologist : he sees the initial physico-chemical phenomenon and the final physico-chemical phenomenon, both of which are recorded by his instruments ; but the intermediate phenomenon, which is the real generator of all that he sees, escapes him completely and the way in which these phenomena combine and are inter-related remains an enigma, although he sees that the fundamental laws of energetics and physico-chemistry have been observed.

But if the engineer and the chemist from Mars possessed the power of reasoning they would see that the hypothesis of an electrical energy endowed with certain clearly defined energetic properties fully accounts for the phenomena observed, that in other words it *explains* them, enables them to be deduced and their most characteristic manifestations to be forecast. They would then be ready to accept this special form of energy, even though they had no instruments for registering it. Why should not the physiologist behave in the same way towards the special form of vital energy which also possesses clearly defined energetic properties, when the hypothesis is able to account for the most fundamental and characteristic phenomena of life, that is for the purposive manifestations, even though he has not yet the instruments which enable him to verify its existence ?

IV

It is animistic vitalism and not energetic vitalism which discourages scientific investigation

The well-known accusation which mechanists quite legitimately bring against animistic vitalists is that they discourage scientific investigation. " Although mechanism," writes Conklin, " may not in the last analysis explain vital phenomena, it is evident that very much of a mechanistic nature remains to be discovered in organisms, and the great advantage of mechanism over vitalism is not only that it is more intelligible, but also that it encourages scientific investigation, whereas a thoroughgoing belief in vitalism discourages research."[1]

" Mechanism," writes Piéron in his turn, " is a vigorous investigating agency, a motor, vitalism a hesitant controlling agency, a brake."[2]

Rabaud is even more emphatic : " Vitalist reasoning is always departing from scientific reasoning which repre-

[1] Conklin, *op. cit.*, *Problems of Organic Adaptation*, pp. 356–357.
[2] Piéron, *art. cit.*, " Du rôle et de la signification du conflit scientifique entre mécanisme et vitalisme," p. 125.

sents a source of action."—" Though the man of science sees the complexity of the field which he is exploring he does not allow himself to be disheartened by his task, but pushes on into the unknown, like a pioneer in a virgin forest. The vitalist, however, is content to believe: his faith satisfies him. Evading all work and effort he finds gratification in imagining a world full of mysterious 'forces' which give birth to miracles. The case of Driesch illustrates this: he was an excellent observer who could have aided the progress of science, but one day he gave in to thie dfficulties of his task; then the entelechy appeared to him in a dream; he thought he saw in it the solution which would release him from further efforts and he seized upon it. From that time he has believed; very soon he ceased to work: his entelechy perhaps has brought him peace, but it has closed his mind to the advances of science, it has sterilized it and prevented him from contributing to the growth of knowledge."[1]

To these accusations, which are undoubtedly justified, the vitalists might very properly reply that the mechanist himself is often content to believe and that his faith not only makes him a little too self-satisfied when he ought to be considering the complete insufficiency of his supposed explanations, but also (the cases of Loeb and Rabaud being typical in this respect) blinds his mind to such a degree that the most evident facts are hidden from him and he is led to deny even those which are in front of his eyes and, in particular, the purposive manifestations of life. "The mechanist disciples of Loeb," writes Piéron, " distort the facts in their attempt to demonstrate the simplicity of the immediate physico-chemical mechanism of a tropism, just as believers in an animal spiritual

[1] E. Rabaud, " Le vitalisme et la science," *Scientia*, March, 1923, pp. 202–203, 230–234.

quality give us improbable accounts of happenings which they say they have witnessed."[1]

Not only his disciples, however, but Loeb himself proves, with his over-simplified explanations of the behaviour of lower organisms based on the theory of tropisms, to what a degree of complacent mental blindness mechanism can lead, and how completely it can blunt that critical spirit which is so necessary if the experiments of the laboratory are to be made really fruitful. In this respect it yields nothing to the most mystical form of vitalism. And again it is Loeb himself who shows us that even the term 'tropism,' which he uses and abuses so much, is nothing but a *flatus vocis* to be placed on the same level as the 'vital impulse' of Bergson and the 'entelechy' of Driesch.

The perfectly legitimate charge brought against animistic vitalism, that it exercises a deadening and drugging influence on the spirit of investigation, does not affect our energetic vitalism, which, on the contrary, by its clear and precise hypothesis of a special form of vital energy endowed with well-defined elementary properties, encourages numerous lines of investigation and experimentation. We shall confine ourselves to mentioning only the chief of these, and especially those which are directed towards controlling, either by verification or refutation, the equally clear and precise deductions which follow from this hypothesis.

V

A new series of experiments suggested by our hypothesis

We shall confine ourselves to discussing here as briefly as possible some of the most typical experiments suggested by our conception of life, systematically arranged in

[1] Piéron, *art. cit.*, p. 119.

connection with the various classes of biological and psychical phenomena to which it applies.

Thus, for example, our explanation of the most elementary phenomena of life, such as assimilation and metabolism, is of a kind to stimulate attempts towards ascertaining and probably to encourage new ways of ascertaining, whether assimilation is really a selective process and how the choice of the particular radicals or atomic groups, which have to build up again the specific part of the living substance destroyed, takes place ; similarly it prompts us to establish definitely whether metabolism is really an energetic process in a stationary state, that is whether all its manifestations (consumption of given substances and the production of waste matter) are constant in time.

In addition, according to our hypothesis metabolism is a process of a vibratory nature ; this can be verified by means of experiments similar to those performed by Engelmann, who has shown that the colours of the spectrum which are the most completely absorbed by certain bacteria are those which are most favourable to their metabolism. Attempts could then be made to discover whether each specificity of metabolism has a vibratory period of its own.

Our hypothesis of a central zone of development could lead to a whole series of experiments on the development of multi-cellular organisms, similar in certain respects to those of cutting infusoria into pieces, which have proved that the nucleus acts as a true central zone of development in all unicellular animals. These experiments would have to be conducted in a way which would prove either the existence or the non-existence of such a central zone of development and then, if a positive result were attained, it would have to be ascertained whether the place which we have assigned to this zone is in accordance with facts. To this category of experiments belong, for example, those on the influence exerted by the nervous system, or by certain of its parts, on generation and regeneration (recorded in several numbers of the *Archiv für Entwicklungsmechanik der Organismen*), those of Speman on tritons and on certain star-fish, which have proved the existence of the true central zone of development in these organisms, and many similar ones, which might for the future be more systematically directed towards this particular purpose of verification.

Again, according to our centro-epigenetic hypothesis, ontogenetic stimulation is simply the restitution or reproduction, by internal causes, of the functional stimulation which, in the past, exerted the same plasmatic action which the ontogenetic stimulation now exercises. In connection with this question we remember that numerous observations and experiments which have already been made argue in favour of a gradual and smooth passage from the embryonic period to the functional period (to use Roux's terminology), that is, in favour of an imperceptibly gradual substitution of functional stimulation for ontogenetic stimulation. We are thus led to assume that during a certain period of development the two stimulations acted simultaneously and together, *as if they were of the same nature.* But innumerable other experiments can and ought to be made for the same purpose, to show, that is, the complete identity of the two stimulations.[1]

This centro-epigenetic hypothesis (according to which the central zone of development is the true place of the origin and the emission of the germinal substance which is subsequently received and further elaborated by the cells of the sexual organs which, as a result of that reception,

[1] See Author's work, *On the Inheritance of Acquired Characters*, London' 1911, Chap. IV.

are changed from somatic into sexual cells) is a stimulus to investigation into the actual origin of germinal substance and into the ways by which the latter is carried from the central zone of development to the sexual organs ; it might also encourage further and more accurate observation of the process of synapsis, which takes place at the moment of what we call the ' maturation ' of the egg and the spermatozoon, and so enable us to decide whether it can be interpreted as a readjustment of the germinal substance in the cells of the sexual organ, which, out of so many others, have by chance received it and hence arrived at ' maturation,' that is, been changed from somatic into sexual cells.

Our conception of the germinal substance as a columnar stratification along the axis of the chromosomes of successive specific accumulations, each of which represents a given phylogenetic stage and at the same time determines the corresponding ontogenetic stage, has already been substantiated by the fact that the chromosomes often appear to be composed of superimposed discs, rather like a necklace, or, more nearly, like the first columnar pile of Volta. But further observation might lead to fresh evidence on this point.

Our suggested explanation of the purposiveness of the phenomena of pre-established adaptation, in which these phenomena are interpreted as the result of the hereditary transmission of new adaptations, is of a kind to lead to new experiments on the transmissibility of acquired characters, different from previous ones in that they would have to aim, first of all, at producing *very clear and localized phenomena of new adaptation*. One experiment, for instance, which we have already suggested to a number of eminent biologists, would consist in fixing by means of pins two very light aluminium rings to any two vertebræ (always the same in successive experiments) of a rat's or a guinea-pig's tail, and then placing between the two rings a spring which would tend to stretch them apart or bring them together. After this contrivance had been fixed the animal could again be set free. It would then remain to be seen whether this external influence, present throughout the whole of the animal's life and tending to lengthen or shorten its tail, would produce this effect or not. Then, repeating this experiment upon a long series of generations, it could be seen whether in the last generation animals would be obtained which, although the mechanical influence in question had ceased, possessed on an average a longer or shorter tail in comparison with the normal measurement of the species.

Another experiment of the same type would consist in bringing a constant pressure to bear on a given point of the skull by means of a suitable aluminium cap fixed permanently on the animal's head, and discovering, as in the previous case, whether, when the experiment had been repeated over a long series of generations, the new morphological adaptation resulting were transmitted in whole or in part to the last generation, which would be freed from all pressure.

Experiments of this kind—producing, we repeat, new morphological adaptations *of a clear and strictly local nature and admitting of exact measurement*—are the only sort which can constitute the true *experimentum crucis*, capable of giving a definitive solution to the problem of the transmissibility or non-transmissibility of acquired characters. On the contrary, this problem will never be solved so long as the new adaptations to be transmitted, instead of being *well-defined*, *very clear*, *strictly localized and exactly measurable*, are undefined physiological modifications of a general order, which may be produced at the same time in all the cells of the organism and therefore in the germinal cells also, which would thus transmit their own modifications and not those of the soma.

Again our hypothesis concerning the nature of new adaptation suggests new experiments which, so far as we know, have not yet been tried. Does every phenomenon of new adaptation in response to an unaccustomed external agent acting on the organism begin, as we suppose, with a disturbance of the stationary state, that is to say with the appearance of a succession of dissimilative and assimilative phases which are all different from each other, instead of being similar as they are when the metabolism is in a stationary condition ? And does new adaptation consist, as our hypothesis assumes, either in a return to the preceding stationary condition, or in the establishment of a new one, as a result of the chance production of a particular substance which is able to neutralize the disturbing agent or involve it in the new metabolic process ? Such are the questions which can be settled by exact investigation, not easy, it is true, and requiring careful handling, but still possible. Even if this investigation might only bear for the moment on the variance or non-variance in time, during the phases of non-stationary unrest and stationary rest respectively, of the consumption or production of certain substances, yet the proof or exhibition of these phases which would thus follow, the verification of the longer or shorter duration of the non-stationary phase of unrest and the constant final appearance of the other phase of metabolic stability, would be of the highest importance and would argue strongly in support of our hypothesis—especially if later experiments should succeed in demonstrating the presence of a feverish temperature during all states of non-stationary unrest, for this would be a confirmation of the mainly dissimilative nature of this state.

After the well-known and accurate observations and experiments of Jennings little remains to be done in order to demonstrate purposiveness of the behaviour of the lower organisms. Any, however, which might serve to prove, *ad abundantiam*, the inadmissibility of Loeb's explanation based on the theory of tropisms would not be useless. Still more useful would be any which brought into greater prominence the phenomena of memory, the capacity for profiting by experience, and the presence of affective preferences, even in unicellular and the lowest multicellular organisms although they have no nervous system, as Jennings has already irrefutably shown.

Experiments designed to bring out the phenomena of memory, learning and affective preference *in vegetable life* would be exceedingly valuable. We will only say here that certain experiments on the persistence, after the re-establishment of normal conditions, of certain new rhythms in plants produced by an artificial alternation of light and darkness, have already proved the existence of phenomena of memory in the vegetable kingdom; and that in our essay on the mnemonic origin and nature of the affective tendencies we were able to quote experiments demonstrating the production in elementary vegetable organisms of new affective tendencies following upon a permanent change in their environment. Other very interesting experiments on the 'nervous' behaviour of plants are the well known ones of Sir Francis Darwin, Prof. G. Haberlandt, and the Indian scientist Sir Jagandis Chunder Bose.

Our theory concerning the reflexes and instincts is of a kind to encourage experiments making still clearer both their mnemonic nature and the analogy between the way in which they are released and the mnemonic evocation of ideas. Time might very profitably be spent in multiplying the very suggestive instances relating to the reflexes and instincts, which it is to Semon's great credit to have brought together, in order to prove their essentially mnemonic nature, so as to build up an imposing array of facts which would open the eyes even of those who obstinately keep them closed.

In its turn, our hypothesis relating to the mnemonic origin and nature of the affective tendencies suggests a series of experiments with the object of stimulating artificially in animals (and even in plants) new affective tendencies, new preferences and new choices in addition to or as substitute for their innate tendencies, preferences and choices.

It also suggests a whole new series of experiments with reference to the transmissibility of acquired characters. Attempts might be made, for instance, gradually to accustom an animal to new conditions of environment in general or to new conditions of diet in particular. Thus we might try to make a particular caterpillar prefer the leaves of certain plants rather than those which it and its species normally eat ; or again, we might try to accustom particular herbivorous animals to a meat diet or carnivorous animals to a vegetable diet, and so on. If our theory is correct the animals would end by preferring the new environment or the new diet to the old. If then these experiments were continued over a long series of generations, the new environmental conditions and the new diets ought eventually, if the transmissibility of acquired characters really exists, to be *spontaneously* preferred, that is without any preliminary individual learning, by the distant descendants of these animals.

And, moreover, supposing our hypothesis as to the mnemonic origin and nature of the affective tendencies to be correct, there flows from it a whole series of analogies between affective evocations and sensorial evocations, between affective successions and inhibitions and successions and inhibitions of ideas, which might be brought to clearer view by objective experiment and introspection. The emotional effects of music and the very nature of this latter as a 'releaser' of feeling might receive new light from these analogies.

According to our theory of specific accumulation as applied to sensory evocations, every memory of a sensation is only the reproduction, thanks to the discharge of the corresponding specific accumulation, of the same specific nervous current which in the past deposited this specific accumulation and which now, as current of discharge, proceeds in a direction opposite to that of the corresponding current of charge. Now all this is capable of experimental verification. This verification, moreover, has already been carried out, as we have mentioned elsewhere in a previous work, in Wundt's well-known experiment (in which the calling up of a vivid memory of a given colour, whilst the observer gazes fixedly at a white surface or a white figure, produces in the object gazed at the appearance of the complementary colour) and by other similar experiments suggested by it. Experiments of the same kind could be devised and carried out in other spheres of physiological activity besides that of sight.

The specificity of nervous currents which we have assumed (and it will be remembered that, according to our hypothesis, each nervous current is made up of a number of 'nervions' of equal energetic capacity, different from that of the 'nervions' of other currents of different specificity) affords us an explanation, as we have shown in previous works, of the mechanism of the association of ideas and inhibition. This mechanism is similar to that which we should have in a network of distribution of electrical energy if in the knots of its meshes were placed electric accumulators, each able to produce—by an hypothesis not actually realizable—an electric current *only of a determined intensity.* Each of these electrical accumulators, even if it were constantly inserted in the circuit, could discharge itself only when the electro-energetic conditions of its immediate environment were such as to allow the discharge at this point of that intensity of current (this would correspond to *psychic association*); whilst it would not discharge itself, even if inserted in the circuit, when the intensity of current, allowed to exist in this point by the electro-energetic conditions of the immediate environmental portion of the circuit, were different from that unique intensity which, by hypothesis, this accumulator would be able to give (this would correspond to *psychic inhibition*).

This hypothesis, as we have seen, explains all the modalities of the phenomenon of association or succession of ideas in which inhibition plays so great a part, and, inversely, these modalities fully confirm our hypothesis. But if this phenomenon of mental association and inhibition is further examined with the precise object of verifying the mechanism which we assume to be at its basis, new and more exact confirmation might be obtained.

Other consequences following from this same hypothesis can be submitted to the control of experiment and observation. Thus the assumption that the specificity of different nervous currents consists solely in the specific energetic capacity of their constituent elements suggests the possibility that when a current releases instead of inhibiting another current, it may strengthen or, to use the term which Sherrington applies to reflexes, 'impinge on' the current which it has released; in other words, it is possible that the neuro-dynamical energy of one of the currents serves to augment the neuro-dynamical energy of the other : " The relations of mutual dependence existing between the nervous elements," says Wundt, " which are such that the nervous fibres play at one and the same time the *rôle* of conductors of excitations and that of distributors of energetic forces, cause it to happen that whenever an activation of energy takes place in any part of this system, *the quantity of this energy is determined not only by its own store but also by that of the neighbouring elements which also produce energy.*"[1]

Ostwald himself observes that the quantity of energy released always depends not only on the quantity of energy stored and capable of being freed (" von dem Energievorrath der zur Auslösung bereit ist ") but also on the quantity of releasing energy (" von dem Betrage der auslösenden Nervenenergie ").[2]

When we take into consideration these " reciprocal neuro-dynamical influences " (" Neurodynamischen Wechselwirkungen "), as Wundt calls them, that is the supplementary energy coming from the releasing nervous accumulations and serving to increase a given quantity of released energy, we can begin to understand, for example, the extreme vividness of certain sensory evocations in hallucinations. " In somnambulism properly so-called," writes Wundt, " it is unquestionably a case of an exaggerated excitability of the sensory centres which gives images an hallucinatory character. We can, without much fear of being mistaken, attribute the cause of this exaggerated excitability to reciprocal neuro-dynamic influences."[3]

Thus we saw that in order to explain certain modalities

[1] W. Wundt, *Grundzüge der physiologischen Psychologie* (fifth edn.), Vol. III, Leipzig, 1903, p. 660.

[2] W. Ostwald, *Vorlesungen über Naturphilosophie* (third edn.), Leipzig, 1905, p. 355–356, 426, *et seq.*

[3] W. Wundt, *ibid.*, p. 668

and effects of attention we had to assume a real and effective 'reinforcement' of the sensory evocation by the affective tendency which released it; the amount of this reinforcement being in proportion to the intensity of the affective tendency.[1]

By placing nervous energy at the basis of all the phenomena of life we obtain an explanation of the fact that the nervous energy of the motor or voluntary centres does not confine itself to 'releasing' the specific energy stored in the muscular fibres, but also contributes in many cases to strengthening, by its addition, the muscular effects themselves. This we see exemplified in experiments in which the intervention of cerebral innervation, that is of a great voluntary effort, always succeeds in bringing about a new and lasting contraction even of a muscle worn out or exhausted by a prolonged faradaic excitation.[2]

We see this again in certain hysterical subjects who, by an enormous output of nervous rather than muscular energy, succeed in bending the iron bars on their beds. "It is incontestable," Morel writes, "that under the influence of nervous excitation there is a tremendous increase of muscular strength."—"I have seen," writes Pinel, "examples of the development of an almost miraculous strength, when the strongest fetters yielded before the efforts of a maniac with an ease which was even more astonishing than the degree of the resistance overcome."—"There is a tremendous development of muscular strength in some maniacs," says Esquirol; "we have seen some of them carrying the heaviest weights, bursting through the strongest bonds, and throwing down several men who were

[1] E. Rignano, *Psychology of Reasoning*, London (Int. Lib. Psych.), 1923, Chap. III.

[2] E. Tanzi and E. Lugaro, *Malattie mentali* (second edn.), Milan, 1914, Vol. I, p. 576.

trying to hold them."—" Under the influence,therefore, of certain modifications in the condition of the nervous system of a sick person," Morel concludes, " a great stimulation (as a result of the discharge of large quantities of nervous energy) and a considerable increase of muscular strength may develop in him."[1]

From these and similar instances we can see what a number of new and interesting experiments and observations might arise from these exchanges of supplementary supplies of neuro-dynamical energy which, far from being confined to the impingements of reflexes upon one another (so ably studied by Sherrington), can be extended, thanks to our hypothesis which makes nervous energy the basis of life, to all vital activities whatever their physiological differences.

Hering himself, moreover, gives us a most instructive example of the way in which observation and experimentation, when brought to bear on the behaviour of the nervous manifestations of man and supplemented by introspection, enable us to venture an hypothesis on the nature and the properties of nervous currents and their specific accumulations; so that our hypothesis, which endeavours to define the nature and the properties of these specific accumulations, cannot fail to become in its turn a stimulus to new observations and new experiments. We know, for example, that he admits that there is " more than one process in the same morphological element and more than one conducting process in the corresponding fibre "; in other words, he admits not only that the various cells of the nervous system may be the seat of qualitatively different nervous excitations, but also that in one and the same cell numerous and various specificities of excitations may be produced, and that each nervous fibre or fibril

[1] A. Morel, *Traité des maladies mentales*, Paris, 1860, pp. 286–287.

may be successively traversed by numerous nervous currents of different specificities.[1]

This is in complete accordance with our hypothesis which recognizes the possibility of the successive depositing in a single nucleus, whether of a somatic or of a nervous cell, of a number of different specific accumulations, and which even considers the nucleus of germinal cells as being made up of a great number of specific accumulations, all different and each representing and determining a definite ontogenetic stage.

The same eminent physiologist goes so far as to affirm that " in a single element of the retina rays of light having different numbers of vibrations can give rise to different processes," a view contrary to that of the school of Helmholtz, which still holds to the " theory of cerebral cells each perceiving only red, or green, or violet."[2]

In our interpretation, these " different processes," stimulated in a single element of the retina, are only so many nervous currents of different specificities, each of which is produced by a corresponding specific ray of light of a given frequency of vibration ; these currents depositing in their respective sensory centres corresponding specific accumulations, which, being subsequently reactivated, produce the mnemonic evocations of these sensations.

For this reason, according to Hering's view, not only is each nervous cell or each neurone, as we have seen, capable of giving several ' Regungen,' that is, several specifically different excitations, but in each neurone new specific energies can, during the life of an individual, be added to the inherited specific energies. And these specific energies, both those inherited and those acquired during life, subse-

[1] Hering, *Zur Theorie der Nerventhätigkeit*, pp. 119–120 ; *Zur Theorie der Vorgänge in der lebendigen Substanz*, p. 82.

[2] Hering, *Zur Theorie der Nerventhätigkeit*, pp. 119, 120.

quently activate each other, according to the specificity of the other specific excitations which at a given moment are active in the nervous system.[1]

This again fits in with our theory that continually new specific accumulations, all different from one another, may be progressively deposited in each neurone, and form, by their reactivation, the corresponding 'Regungen' of the neurone. It also confirms our explanation of the association of ideas and of all other nervous and physiological activities in general, or of their reciprocal inhibition, as being dependent on the particular neuro-dynamical condition which is dominant in the immediate neighbourhood of each particular specific accumulation.

Hering's theories as to the ways in which nervous energy functions and behaves were the fruits of continued observation and experimentation. That when taken as a whole they are so nearly in agreement with those which we have deduced directly from our hypothesis of specific accumulation (which Hering, however, did not know), is proof of the fact that our deductions are eminently suited for experimental verification, if it is carried out in the spirit of reasoned criticism which can alone give significance to experiments and make them reveal the secrets of nature.

Finally, this hypothesis which postulates nervous energy as the basis of life, is rejected, as we have seen, by mechanistic physiologists simply because the instruments they have can only register physico-chemical energies. It ought, therefore, to stimulate those physiologists who are fully aware of the complete inadequacy of their instruments when the most profound manifestations of life are to be examined, to try to find new instruments suitable for true and proper biological investigation, that is to employ living organisms or surviving parts of dead organisms which

[1] Hering, *Zur Theorie der Nerventhätigkeit*, pp. 121, 123–124, 131.

have been separated or isolated in some way and which ought to be rendered capable of revealing the strictly vital manifestations which inevitably escape instruments designed merely for the recording of temperatures, calories, phenomena of light, electric currents and other similar manifestations of the inorganic world.

In concluding this very-brief account of the experiments which our theory of energetic vitalism might suggest, we can only express the hope that at the very least mechanists will not venture to level against it the criticism which they quite legitimately bring against animistic vitalist theories ; far from having a weakening or deadening effect on the keen spirit of investigation we see that it cannot but stimulate and encourage this spirit to an infinitely greater extent than the mechanistic theory which shuts its eyes to all vital phenomena which are not of a purely and narrowly physico-chemical nature.

VI

Our hypothesis concerning the nature of life stands therefore as a half-way theory which is capable of settling the age-long controversy between vitalists and mechanists.

Even Bayliss, extreme mechanist though he is, admits that this energetic vitalism of ours cannot be criticized on the ground that it drugs the mind of the investigator with empty words and thus deprives him of all stimulus to further research. " Belief in a ' vital principle,' unless so guarded against as to play no part in the *scientific* outlook of an investigator, is clearly dangerous, in that it would tend to bring contentment at a stage in research when further work might result in greater simplification." " But an objection of this kind does not apply to what might be called a mild form of vitalism, in which it is held that there is to be found in living organisms a form of energy not met with in the non-living. It seems to be admitted by those who advocate this view that the particular form of energy in question obeys the laws of energetics and can thus be measured by transformation into heat or other known forms. The fact that we have no instrument by which it can be measured directly is not a serious argument against its existence, since we have no direct means of measuring chemical energy."[1]

To postulate a form of energy different from inorganic forms and to deduce from this hypothesis results admitting of experimental verification is not "to fall back upon mystical tendencies," as Piéron expresses it (an indictment which is perfectly legitimate so long as it is confined to animistic vitalism), but rather to remain faithful to the most strictly scientific and positive spirit; while to formulate this exact hypothesis upon the elementary properties of this new form of energy which is assumed to be at the basis of life is in no way an "abandonment," to recall Rabaud's criticism, nor "a divergence from scientific reasoning which is a source of action." Scientists have been obliged to postulate a new form of energy in the case of amber which, when rubbed, attracts pieces of paper to itself, in order to explain this phenomenon which is so different from anything observed till then in ordinary natural bodies. Rabaud would, I trust, admit that this hypothesis, far from representing a surrender, has become a new and an extremely fruitful source of activity. How can we then, on scientific grounds, deny the necessity for an analogous procedure in order to explain an essential property, the purposive property, which, as we have seen, is common to a clearly defined category of natural

[1] Bayliss, *art. cit.*, "Vitalism," pp. 292, 293, 294.

bodies and, on the other hand, is absent in lifeless bodies ?

The assertion made by Anthony concerning vitalism in general, namely that it ends by abstracting living matter from the principle of causality, likewise does not apply to our energetic vitalism. " It is easy to perceive," writes this author, " that vitalism can in no way be fitted into the definition of science as a causal explanation of the universe. It always ends by withdrawing living matter from the principle of causality or from its immediate and logical consequences."[1]

Far from meriting this criticism, our hypothesis succeeds, on the contrary, in giving a causal and deterministic explanation, solely by means of forces *a tergo*, of all the purposive manifestations of life, by substituting for ' final causes,' which are inadmissible in science, actual causes which hold throughout the rest of nature

Far from withdrawing vital phenomena from determinism, our conception of life, because of its theory of the mnemonic origin and nature of all the ' tendencies ' of the organism, whether organic or affective or intellectual, extends its validity to the whole biological and psychological sphere, so that it covers even those cases in which the facts might seem to be at variance with determinism : as when, for instance, ontogenetic development or physiological reactions or animal behaviour or man's conduct present differences from one case to another although all the conditions of the phenomenon appear identical. Our theory gives an unqualified explanation of all these cases by taking into account the great number and variety of individual mnemonic accumulations, especially those of an affective character, which makes the slightest external or

[1] R. Anthony, *art. cit.*, " Sur le sens et la portée du vitalisme," p. 398.

internal differences sufficient to activate sometimes the one and sometimes the other.

Nor in our energetic vitalism is there any *opposition* between vital laws and mechanical, physical and chemical laws. This opposition is present, however, in the ideas prevalent among animistic vitalists: "The principle of vitalism," Gley writes, "kept its force even when the XIXth century was well advanced. For Müller and Liebig, as for Cuvier and many others, it is quite possible that living bodies are under the influence of the same physico-chemical forces as those which are active in inorganic nature, but there exists a force by which the processes which go on in these bodies receive a different impulsion from that which they would have received outside these organisms; this force knows all the secrets of physical and chemical forces and it acts *in conflict with them* as the cause and regulator of all the phenomena of the living being."[1]

Nothing of this is applicable to energetic vitalism, of which we can simply say that when, in accordance with the laws of energetics, the inorganic forms of energy making up the external nutritive, stimulative environment are transformed into the form of energy peculiar to life, the latter from that moment follows its own laws, just as the electrical energy, into which the mechanical energy from water flowing into the turbines is changed, obeys its own laws.

Thus although the accusations of being at variance with physico-chemical laws, with determinism, with the principle of causality and the basic laws of energetics, and of assuming for life concepts which oppose and contradict the most incontestable truths of science, have been justi-

[1] E. Gley, "Le vitalisme. Evolution de la question," *Scientia* July–August, 1923, p. 16.

fiably levelled against animistic or mystical vitalist theories (into which category that of Driesch with its mysterious and nebulous entelechy unquestionably falls), yet they do not in any way affect our vitalistico-energetic hypothesis. If a full realization of the anti-scientific nature of any attempt to explain the indisputable purposive manifestations of life by the doctrine of 'final causes' is reached, then it will be seen that the introduction by our energetic vitalism of a special form of energy with well-defined elementary properties is made precisely in order that we may give a strictly causal and deterministic explanation of the purposiveness of life—a thing which is impossible so long as we adhere to physico-chemical energies alone.

Our hypothesis, therefore, escapes both the criticisms which the vitalists quite properly bring against the mechanists when they say that the latter cannot account for the most fundamental manifestations of life, and those which the mechanists just as properly bring against the vitalists when they say that their metaphysical entities, which are mere verbal expressions, do not explain anything. This hypothesis of ours springs, indeed, from a recognition of the justice of the criticisms of both sides, and is an attempt to meet both of them.

It represents consequently an attempt at reconciliation, an intermediate theory which is capable of bringing to a conclusion the age-long controversy between vitalists and mechanists.

CHAPTER XII

CONCLUSION

MAN'S PLACE IN THE UNIVERSE AND THE NEW MORALITY OF THE HARMONY OF LIFE

An impassable gulf separates life from the inorganic world. Non-purposiveness of the Universe as a whole as opposed to biological purposiveness. It is utterly useless, therefore, to look for the teleologism or the end or the purpose of the Universe itself. We can, on the other hand, profitably and with some hopes of success, look for the teleologism or the end or the purpose of life. Man, who in origin is united with all the rest of life, becomes more and more completely separated from it since only in him—as a result of social life and its two supreme ends, justice and morality—does the harmony of life take the place of the struggle for life. Harmony of life thus appears to be the ultimate and supreme end of the biological portion of the Universe, which alone shows itself capable of purpose. Though our conception is not one which will satisfy religious people, who are impelled from their inmost being, either as a result of innate mysticism or acquired mental habit, to believe in the divinity and in the immortality of the soul, it can, on the other hand, satisfy all those idealists who even in their striving towards the loftiest ideal refuse to sacrifice reason to sentiment. They see in the joyous harmony of the whole of life the true undying ideal, which is the *raison d'être*, the supreme end, the animating motive of their transitory individual existence.

IF, having come so far, we wish now to trace in broad outline the conception of the Universe or the *Weltanschauung* which emerges from our biological synthesis, we find, in the first place, that it draws a distinct impassable line between life and the whole of the inorganic world. Assuming that all the purposive manifestations of life depend entirely and exclusively on the mnemonic property, that is on the fact that it is in life and only in life that a particular form of energy is active, the particular energetic characteristics of which are the sole cause of these purposive manifestations, then it clearly follows that to the biological purposiveness which results from this mnemonic property there must be opposed a non-purposiveness of the inorganic world, that is of the Universe as a whole.

We can thus realize the inevitable futility of every attempt, from the first mystical and uncertain gropings of philosophical speculation to the great metaphysical systems of the greatest thinkers of past centuries, to discover the goal towards which the Universe is tending, or the mysterious design of the supreme divine master when he created the Universe and imparted to it constant and eternal movement. A fruitless search, indeed, because the Universe as a whole is not teleological, it has no purpose in view, it tends towards no end, for the simple reason that the particular form of energy which is the sole source of every purposive manifestation is active only in its biological section.

But though it is utterly useless to look for the teleologism or the end or the purpose of the Universe as a whole, we can profitably and with the greatest hopes of success look for the teleologism or the end of the purpose of that part of the Universe constituted by life, which alone is animated by this particular form of energy which imbues with purposiveness every one of its manifestations.

From the first uncertain and tentative efforts of unformed primordial protoplasm to the most complex and perfect vertebrates the biological finalism which we see first at work is less a single finalism of life as a whole than a continual and chaotic outburst of particular finalisms clashing violently one against another and destroying each other. This results from the fact that in this first period of organic evolution each organism, because of its own narrow finalism, strives only for its own conservation in opposition to and, if need be, at the expense of all the other organic units, which at the same time are striving just as eagerly for the realization of their own particular purposes.

Then from this unceasing and violent chaotic clash of individual finalisms co-ordination was slowly and gradually evolved as a result of the development of increasingly complex organic units in which countless myriads of elementary vital pulsations came into accord and began to supplement each other. Whenever two particular finalisms instead of clashing began, even though in a purely fortuitous fashion, to harmonize, then *ipso facto* they worked together to maintain and perfect this harmony, simply because this latter was satisfying to both of them. At the same time the harmony thus established between the two elements, acting as a conserving and consolidating factor of both their lives, constituted for them an element of success in the general struggle for life.

Finally, with the origin of man, with the appearance and development of this very lord of life, the slow and gradual co-ordination of single chaotic finalisms, culminating in the formation of harmonious organic wholes, passed from the realm of simple biological individuality, in which it manifested itself in the creation of more and more complex organisms, into that of collective life. Just as the particular finalisms of several cells tended to harmonize and blend, whenever chance made this possible, and so gave birth to the first multicellular organisms and then to more and more complex and perfect multicellulates, so the particular finalisms of our most remote ancestors began sporadically to converge whenever the possibility of substituting harmony for conflict and clash appeared. Here again this was achieved as a result of the individual purpose of self-conservation, which impelled towards the creation of a social organism as the best means of satisfying this purpose ; and from the primitive patriarchal family, the tribe and the clan to our modern societies social life has continually moved towards greater extension and complexity, as well as towards completer co-ordination and fellowship.

Thus man, who was originally one with the rest of life, gradually became separated from it, because in him alone—thanks to his ever wider and more complex collective life and to the stimulation of an ever more intimate spiritual communion through language and other inventions of communal life—have arisen and developed, at first unconsciously but subsequently more and more consciously, the two great social ideals of justice and morality ; ideals which are tending slowly and painfully but with increasing success to substitute harmony of life for the struggle for life.

Harmony of life, the end which man so ardently desires, thus stands out as the ultimate and supreme purpose of the biological portion of the Universe, which alone is really purposive.

If we now consider the relation of our conception of life to the essentially antagonistic mystico-spiritualist and materialistico-positivist conceptions, which for centuries have struggled and still are struggling for domination, we shall see clearly that although it is unequivocally anti-animistic, it does not shock, as the mechanist theory does, the promptings of our innermost sentiments which will not allow us to believe that man is a mere machine, a mere physico-chemical process of the same kind as those which take place in our laboratories.

The feeling that he is attached by his innermost nature to life as a whole, and, on the other hand, that he is essentially different and distinct with the whole of life from the inorganic world, will lend fresh stimulus to man's sympathy with all living beings, and more especially with the representatives of his own species, in whom life pulsates in closest unison with his own. On the contrary, the belief that he is nothing more than one of innumerable insignificant and fortuitous links in an unending and purposeless physico-chemical transformation, on the same level as brute matter, must inevitably have a depressing and enervating influence on man, crushing all aspiration towards advancement and moral elevation. In place of this we substitute the optimistic view, infinitely more reassuring and encouraging, that man is the supreme product of an age-long and unremitting self-elaboration of that powerful and wonderful cosmic activity which is life, which, thanks to its mnemonic property, stands alone in its capacity for progress, and for immense progress such as that which has been achieved from the first tentative efforts of its infinitely distant and humble origins to the appearance of the crowning spirit of genius and the noblest ideals of the morally finest men.

Religious persons—for whom we must all naturally feel a profound respect—can find, it is true, in consequence of their innate mysticism or their education or social 'nurture,' no comfort in life nor support for their conduct other than that furnished by faith and trust in the eternal survival of the individual soul beyond the mortal body; they will, of course, be no less hostile to our conception of life than to the mechanistic. Apart from these mystical persons, however, in whom feeling reigns tyrant over reason, this religious need is not innate in the normal man. As we have shown in other works, it results from a sentimental tendency which, in view of very necessary social ends, it was the business of the church, in the first stages of collective life, to implant in all the members of the community. It is, therefore, acquired and transitory and tends in an increasing number of members of the community to disappear and yield place to a positivist conception of the Universe, according as the primitive social needs, for which religion was created, lose their *raison d'être* in our civilized societies. We see this illustrated in the growing atrophy which seizes upon the religious organ, when it does not gradually transform itself, as among the Anglo-Saxon peoples, into a purely educational and moral institution.

Thus a new religious attitude (if we wish to retain this term for the designation of all idealistic aspiration) is tending to take the place of the old, assuming new forms which are in full accord with science, enlarging its boundaries, and placing its ideal ever higher and higher. The man whose mental and emotional attitude has undergone this change will find much to satisfy his soul in recognizing himself as a small but intensely vibrating part of the vast current of life which is truly immortal and whose advances in the past are the certain warrant of an indefinite progress in the future.

Narrow and egotistical aspiration towards the immortality of the individual soul will then yield place to a loftier and more altruistic aspiration towards the completer and more harmonious fusion of his own with the universal rhythm of the whole sphere of life. The individual will then willingly accept as a noble and eagerly coveted privilege the duty of diminishing sorrow and increasing joy among all his kind to whom he is so intimately attuned. And he, too, will aspire to immortality, but to a special and very lofty immortality: that is he will fashion his life so that even after material death he will continue to be for all the rest of life an ever active factor towards progress and good. This will be possible of realization thanks to the biologico-mnemonic property which is transformed within society into social memory.

And it will be not merely the great and immortal works of science and the arts which will bring solace and joy to suffering humanity in even the most distant future ; not merely the gigantic and celebrated achievements of technical and economic activity which will be able to boast of having realized by their adaptations of the external world to human life a benefit which will be felt even in the most distant future epochs ; not merely the sublime words of a Socrates or a Jesus of Nazareth which future generations will continue to bless until the end of time for having indelibly imprinted in the human mind new moral postulates which are more in harmony with the supreme end of life ; but every action, however insignificant it may be, performed with a view to good, every act, however small, inspired by justice and love, will leave its imperishable mark on the coming race, and pass on with an ever beneficent influence from one individual to another into the remotest ages.

Hence it is in vital purposiveness itself which reaches its culminating point in society in the supreme principles and lofty ideals of justice and morality, it is in his own works which aim at contributing in the fullest measure possible towards the coming of a nobler and finer human life, that the man of to-day, imbued with this new positive idealism, will find the means of satisfying his aspirations, which are no less profound, no less irresistible than those of the most profoundly religious man.

And it is in the joyful harmony of all life that he will see the true undying ideal which is the *raison d'être*, the supreme purpose, the animating motive of his individual ephemeral existence.

INDEX

www.ingramcontent.com/pod-product-compliance
Lightning Source LLC
LaVergne TN
LVHW021107300926
849160LV00009B/647
* 9 7 8 0 4 1 5 6 1 4 2 0 7 *